The Giza Park Protocol
How to Start the Sun

Operations Manual #1
Termination of Ice Age

Captain Leo Walton

THE GIZA PARK PROTOCOL
HOW TO START THE SUN

OPERATIONS MANUAL #1
TERMINATION OF ICE AGE

by
Captian Leo Walton

THE GIZA PARK PROTOCOL HOW TO START THE SUN

Copyright © 2025 Captain Leo Walton

All rights reserved.

This book is intended for informational and entertainment purposes only. The content is provided "as is" and reflects the author's personal opinions and interpretations. While every effort has been made to ensure accuracy, the author and publisher make no representations or warranties of any kind, express or implied, regarding the completeness, accuracy, reliability, or suitability of the information contained herein.

The material in this book should not be considered as professional advice (legal, financial, medical, or otherwise). Readers are encouraged to consult with qualified professionals before making any decisions based on the content. The author and publisher disclaim any liability for any loss, damage, or inconvenience arising from the use of this book.

All characters, events, and places described are fictional or used in a fictitious manner, unless explicitly stated otherwise. Any resemblance to actual persons, living or deceased, or actual events is purely coincidental.

No part of this book may be reproduced, distributed, or transmitted in any form or by any means, including photocopying, recording, or other electronic or mechanical methods, without the prior written permission of the author, except in the case of brief quotations embodied in critical reviews and certain other non-commercial uses permitted by copyright law.

Dedication

I dedicate this text and those who continue this research to the necessary expansion of Ancient Pyramid Systems research. We have spent too long believing in a history of lies and manipulations for a particular agenda.

We must break the chains of brainwashing and escape slavery over our minds for the betterment of our planet and the existing human species struggling to survive the pandemic and the latest attacks on our sovereignty.

Understanding these Ancient Systems is the first step in understanding our universe.

All branches of science, including the passionate efforts of individual researchers, need to contribute their collective skills and gifts.

Each contributes their color to create a more complete canvas, an art of the best civilization ever to populate the Earth.

Once you read this first introductory volume, you will recognize your calling and optimize your vision passionately.

If we want to understand the Mysterious "Hall of Records" symbolism, we must know that it is written in stone in Egypt, Bosnia, Teotihuacan, Angkor Watt, China, and more.

Let's begin.

Acknowledgment

My list of Acknowledgements is much longer than I can logically print. I have always claimed that any success I have experienced in my life has been by me standing on the shoulders of these Giant Influencers in my life.

Beginning with my parents, Leo Walton Sr. and Shirley Schumaker provided a loving, supportive, and adventurous childhood environment that encouraged creativity. My grandfather, Charles Schumaker, taught me to be observant and to question everything.

My cousin Richard Shaw was like an older brother who always pushed me to excellence in everything I did.

The list includes the great works of Plato, Herodotus, Hermes Trismegistus, Nikola Tesla, Dr Semir Osmanagich, Chris Dunn, John Cadman, Andrew Collins, Grahm Hancock, Edgar Casey, Mark Lehner, John Anthony West, Robert Shock, William Brown, Neil Helm, Michael Tellinger, Geoffrey Drumm, and Robert Duvall, which are more directly related to this Pyramid project.

More recently, my new friends who have contributed are Kevin Muldoon, Sandra Prieto, Barabara Totsi, and Jesper Johansen.

Finally, I have the awesome support of my soul mate, Heidi Anne Grant, who always has my back, is the most fantastic cruise director, and is ready to go on a well-planned and arranged exciting adventure.

THE GIZA PARK PROTOCOL HOW TO START THE SUN
Table of Contents

Dedication iii
Acknowledgment iv
Chapter One: Personal Introduction 1
Chapter Two: Biography 25
Chapter Three: Earth & Egyptian History 34
Chapter Four: Earth's Cosmic Connection 52
Chapter Five: The Odyssey Begins 69
Chapter Six: Sea to Shining Sea Tour & More 87
Chapter Seven: Return Trip to Montana 167
Chapter Eight: Peru Nazca Lines 184
Chapter Nine: Welcome to Egypt 205
Chapter Ten: The Egypt Taurus Correlation Theory 248
Chapter Eleven: Egypt Weather and Climate 297
   "Terminating the Ice Age." 304
   Part Two 320
Chapter Twelve: The Bosnia Pyramids Pleiades Correlation Theory 391
Chapter Thirteen: The Teotihuacan Aries Correlation Theory .. 412
Chapter Fourteen: Hieroglyph of Ani, Plate # 2 "Sunrise" 416
   Inventory of Components 423
   The Greatest Discovery on the Giza Plateau, Osiris's Tomb 428
   Energy Analysis of the 3rd Level of Osiris's Tomb 434
Chapter Fifteen: Summation overall 440
   The Giza Connection: A Lost Technology 445

THE GIZA PARK PROTOCOL HOW TO START THE SUN
Tracing The Energy to The West Coast 446
Lone Butte—A Hidden Key to the Ice Age Melt 447
The Osiris Line and the Ring of Fire 448
Weather and Climate Modification Systems (WCMS) 450
A Personal Reflection 452
A Final Thought 454

Chapter One:
Personal Introduction

Hello, everyone. I'm Captain Leo Walton, and my previous career was as a successful consultant and a self-employed Marine Forensics Investigator for over 15 years. I represented 19 private and commercial marine insurance companies in the US and Lloyd's of London's Atlantic small boat fleet operating in US waters.

Over the course of 15 years, I conducted more than 2,500 forensic investigations. I was contracted to work with seven different hurricane catastrophe teams, following the paths of the most destructive hurricanes to strike the East Coast of America during that period.

My expertise extends beyond investigations. I have conducted private training sessions for insurance adjusters and authored three unique training manuals. These manuals cover a range of topics, from emergency services for sunken vessels to diagnosing commercial vessel engine failures. They offer innovative, cost-saving solutions that have proven invaluable to the companies I worked with.

My experience is not confined to a single area but spans multiple specialities. I have conducted mechanical failure analysis, chemical laboratory analysis, oil spill containment, recovery of sunken vessels, lightning strike investigations, fire investigations, and injury or wrongful death investigations. I have also been involved in accident investigations and have served as an expert witness in court.

With over 250 forensic lightning strike investigations to my name, I may have conducted more research on lightning damage than almost anyone else in the field.

THE GIZA PARK PROTOCOL HOW TO START THE SUN

My personal experiences with lightning have further deepened my understanding. I have been involved in five lightning strikes, including two instances where current passed through my body and one incident where the force of the strike blew a crown off my tooth.

These experiences have enabled me to develop a unique understanding of the behaviours of plasma energy, which forms the basis of lightning. I came to appreciate Nikola Tesla's fascination with lightning and the reasons he conducted lightning experiments in Colorado.

Before my career in marine forensics, I trained as a nuclear apprentice and contributed to building reactors for the US Navy's nuclear-powered vessels, including the aircraft carriers *Nimitz* and *Vincent*, and the frigates *California*, *Virginia*, and *South Carolina*. I also worked on the final nuclear refuelling of the eight reactors aboard the first nuclear-powered aircraft carrier, the *USS Enterprise*, which served as inspiration for the science fiction series *Star Trek*.

Later, I shifted my focus to submarines, where I was responsible for installing ballast system controls for dock levelling systems and the emergency blow systems aboard the *Los Angeles*-class submarines. This system allowed submarines to perform emergency surface manoeuvres, and I often tell people with a smile, "I made submarines fly!" lol.

The image is Dr Semir Osmanagic introducing me at the 7th Bosnian Pyramids Scientific Conference in 2024.

Taking time away from my Navy work, I transitioned to commercial ship repair, focusing on liquid natural gas container ships, cargo container vessels, and petroleum oil tankers. Over time, I began to move away from my shipbuilding and repair career, which I felt had played a vital role in supporting the protection of my nation and family. I decided instead to contribute more directly to my local community.

THE GIZA PARK PROTOCOL HOW TO START THE SUN

I trained and served as a professional firefighter for two local communities, where I assisted in rescuing accident victims, providing emergency medical services, and conducting fire prevention, suppression, and investigation.

Eventually, I returned to the manufacturing sector, gaining extensive corporate management experience, which led to contracts to establish two companies in Virginia, USA.

The first company was a German manufacturer that required me to undergo two months of training in the UK, alongside a team of 12 new hires. During this time, I authored my first mechanical assembly manuals for escalators and auto-walk human conveying systems used in airports and railway stations. This involved translating technical blueprints and parts inventories from German to English. Before designing, manufacturing, and assembling these systems in the US, I played a key role in setting up the factory and hiring two team members fluent in German to assist with this technical work.

The second company I helped establish involved relocating one of the oldest vertical pump manufacturers in the US from Chicago to Virginia. This process included hiring skilled local tradespeople to fill essential positions.

This second role unexpectedly led to one of the most significant historical contributions of my life. I became the project engineer on

THE GIZA PARK PROTOCOL HOW TO START THE SUN

a contract at the World Trade Center, Twin Towers, in New York City.

I was tasked with the removal, overhaul, and reinstallation of a critical component: the natural gas engine-driven emergency fire suppression pump and its electrical switchgear systems. This pump, located in Building #7, powered the automatic emergency sprinkler systems and emergency standpipe systems, protecting the Twin Towers and their occupants—up to 70,000 people on any given day.

To accomplish this, I negotiated with workers' unions to secure the necessary heavy equipment and contractors. I applied for and obtained permits, arranging for public service and police officers to ensure the safe closure of a large section of Manhattan. Twice, I oversaw the complete shutdown of streets in New York City to enable contractors to remove and reinstall this vital system.

Reflecting on this project, it stands as one of the most significant and lifesaving events of my career.

THE GIZA PARK PROTOCOL HOW TO START THE SUN

This high-pressure turbine pump, capable of pressurising water to 110 storeys, enabled firefighters to save over 60,000 lives in the Twin Towers on 9/11.

A few years later, in 2007, my successful career was abruptly cut short when I was injured in an accident aboard my vessel. I was working on the energised electrical shore power system when the incident occurred.

Following my recovery, I shifted my focus and expertise as a trained forensic investigator from ships and manufacturing to becoming an independent researcher of ancient phenomena. As part of this transition, I engaged in extensive hands-on experimentation and investigations into subtle quantum energies. During this time, I discovered scalar longitudinal waves, which are intrinsically linked to the injuries I sustained from a lightning strike in 1996 and an electric shock 11 years later in 2007.

THE GIZA PARK PROTOCOL HOW TO START THE SUN

Over the past 17 years, my research has been deeply influenced by the lingering effects of these events and the associated frequencies within my body. During the accident in 2007, a 220-volt AC current at 50 amperes arced from an electrical panel I was holding, entering my left hand and coursing through the left hemisphere and frontal lobes of my brain. This caused severe traumatic brain injury (TBI) and damaged both my pituitary and pineal glands, leaving them dysfunctional.

In addition to TBI, comprehensive medical testing revealed several severe conditions: post-concussive syndrome, cognitive disorder, left-side hemiparesis, and chronic fatigue syndrome. These diagnoses profoundly impacted my life.

It took me an entire year to regain the ability to read. Although I could recognise individual words, I could not comprehend their collective meaning; they appeared as an indecipherable foreign language.

The electric shock rendered me unconscious and induced a near-death experience (NDE). I later realised my survival was due to an extraordinary bioenergy event known as a "sudden kundalini awakening," a phenomenon as ancient as humanity itself.

For three weeks, I drifted in and out of consciousness, confined to my bed and entirely alone. For nearly five months after what I call my "grand illumination," I lost all sense of touch. I could not feel the warmth of the sun on my face, the wind in my hair, or even the texture of my own skin. Internally, I felt emotionally numb as well.

Yet, despite the absence of sensation, pain was ever-present. The myelin coating on my nerves, akin to the plastic insulation on wiring, had melted off due to the intense heat generated by the electrical arc. It required temperatures of approximately 2,600°F to melt the stainless steel and magnetic bracelet I had been wearing, which had attracted the electricity.

I believe the small magnet within the bracelet acted as a circuit breaker when it melted, likely saving my life. However, the magnet itself was blown out of the bracelet during the arc.

Five months later, as suddenly as my senses had vanished, they returned with an acute and profound intensity. What initially felt like a miraculous recovery quickly turned into horror.

My body began reacting violently, twisting and contorting in agonising pain as I experienced a strange, invisible current coursing through me. This relentless sensation was triggered by resonance frequency vibrations from my boat's refrigerator system. Each time the reciprocating compressor and electric AC motor cycled on, I felt as though I was being electrocuted all over again.

The refrigerator had become an unrelenting source of torment, turning what should have been a simple appliance into a constant, unwelcome aggressor in my home.

The alien and tormenting physical sensations I experienced at that moment were debilitating. Without realising it at the time, I had been initiated into a new reality. I was embarking on an uncharted journey—one I had neither volunteered for nor agreed to.

Most of the education I have gained since has emerged from over a decade and a half of research, combined with hands-on experimentation using both ancient and modern healing methods. These methods include sungazing, living for 90 days as a breatharian (consuming no food or prescription drugs), radionics, blood electrification, yoga, reiki, deep meditation, brainwave entrainment, astral travel, and consultations with over 40 medical doctors, psychologists, and psychiatrists.

To this list, I can add my abundant on-the-job training (OJT) encounters with paranormal and unnatural experiences. Many of these experiences were driven by a newly discovered skill—a super-intuitive ability to source information visually. This intuitive gift led me to start drawing my visions. Over time, these visions transitioned into words, then into understanding, followed by knowledge. Eventually, I used this confidence to create hypotheses and theories.

For the longest time, the most consistent response I received from the medical professionals I sought help from was that I was a "miracle man." No one knew what to do with me; I was an anomaly who was supposed to be dead.

Even my physician seemed perplexed during my recovery. She conducted physical and lab tests while I engaged in sungazing—staring at the sun with the naked eye. I suspect she thought I had lost my mind, but I eventually gained the ability to stare directly at the sun for 45 minutes. Over the course of 18 months and two sungazing experiments, I spent a total of 153 hours looking at the sun.

These experiments eventually led me to lose my desire to eat. I decided to see how long I could survive on just black coffee and water. On the 75th day of my fast, I went to my physician for a full physical and detailed lab work. During the examination, she asked how I would react if the results revealed less-than-perfect health. I told her that if anything was amiss, I would resume eating immediately.

We had developed an honest and straightforward relationship by that point, and one day she said to me, "Leo, I've been practising medicine for 30 years. In every case involving patients injured like you, I've had to lift the sheet off their faces to identify them. You truly are a miracle man."

When the results of my tests came back on the 90th day of my fast, every result was perfect—except for a 10% deficiency in protein. It was incredible to think that I had gone 90 days without food, with all my tests being perfect except for a slight lack of protein. True to my word, I began eating again. However, since 5 March 2016, I have only eaten one meal per day. I am rarely hungry and can go days without food.

THE GIZA PARK PROTOCOL HOW TO START THE SUN

When I conducted my sungazing and fasting experiments, I was aware that my guru used to go long periods without food but would always consume a small amount of buttermilk. At the time, I dismissed this practice because I wanted my experiments to be entirely free of food products. Looking back now, it's clear why he chose buttermilk—it provided essential protein.

Fast-forward to the present. From the moment I was electrocuted to today, my life has revolved entirely around energy, frequency, and vibration. With limited information available from Western medicine at the time, I turned to alternative methods to regain some measure of physical and emotional stability.

For two years, I was unable to turn my head to the left. Seeking relief, I began training in yoga for its therapeutic benefits. The results were swift and noticeable, providing me with a vehicle to not only heal physically but also immerse myself in Eastern philosophy.

I became deeply committed to yoga and fell in love with both the practice and the community it brought into my life. Through my training as a yogi, I gained access to a wealth of ancient skills and wisdom that proved to be invaluable for my healing—something I so desperately needed at the time.

Despite the profound improvements I achieved through yoga, one area of my life remained unresolved. While my inner bio-energies had become more manageable, they were still beyond my control. These energies would surge through my body, causing extreme physical sensitivities, and seemed to have a will of their own. I was merely an unwilling passenger subjected to their erratic whims.

The aftermath of these episodes left me riddled with anxiety, feelings of defeat, and doubts about my sanity. It wasn't until I began training in the Japanese healing art of Reiki that I found a way forward. By becoming a Reiki Master, I finally developed the skills and confidence needed to regain control over my inner bio-energy fields.

In my relentless pursuit of improvement and adaptation to my heightened sensitivities, I gradually discovered that I could sense industrial and microwave energies, radio fields, as well as natural Earth and cosmic energy streams. Over time, I honed these abilities to the point where I could close my eyes, slowly turn in a circle, and accurately identify all eight directions of the Earth's electromagnetic fields. When tested with a ship's magnetic compass, my accuracy was indistinguishable from that of the instrument itself.

This ability earned me recognition as a Magneto Receiver or Geomancer, given my capacity to detect electromagnetic streams.

Another fascinating ability emerged when I noticed that I could feel the vibrations of ship foghorns in my body before hearing them. This happened frequently near the harbour where I live. On foggy days, when ships from the Atlantic Ocean sounded their deep foghorns, I could predict the sound with such precision that I could tell others when the horn would blare.

This phenomenon led me to theorise that certain sound vibrations operate as longitudinal waves—scalar in nature—and travel faster than the speed of light.

Similarly, I found that I could feel the Sun's telluric vibrational frequencies, associated with its "bow wake," before sunrise. This sensitivity allowed me to tell others the exact moment the Sun had risen, even without seeing it.

I also became attuned to specific microwave antenna systems, sensing and even hearing them when activated. Over time, my abilities expanded to include precognition, visions, and an intuitive capacity to access information when needed. I later identified the source of this information as the Akashic Records—a repository of universal knowledge.

THE GIZA PARK PROTOCOL HOW TO START THE SUN

The transformative experiences in my life following my "Grand Illumination" ignited an obsession to understand brain functions, near-death experiences (NDEs), and their profound after-effects. I became a self-appointed International Ambassador for the International Association for Near-Death Studies (IANDS) and plan to explore parallels between my own experiences and the ancient rituals of Egyptian pharaohs.

By the time you read this volume, four physical books detailing my research and discoveries will have been published. I have written a dozen manuscripts in various stages of production and created hundreds of original drawings based on the visual downloads I received.

From the beginning of my writing journey, I aspired to produce a well-researched, thoughtfully crafted, and coherently written series. Thousands of hours of research and journaling have culminated in new perspectives on a wide range of topics, including the misnamed Egyptian Old Kingdom dynasty, the Giza pyramids, theories on the Ice Age's termination, cultural and natural human cranial deformations, Earth's bio-energy systems, and more.

This work also introduces several groundbreaking theories, such as the Egypt-Taurus Correlation Theory, the Bosnia-Pleiades Correlation Theory, and the Teotihuacan-Aries Correlation Theory. I explore ley lines, Captain's Armillary, and correlations involving the Egyptian and Bosnian pyramids.

Additionally, I propose new interpretations of nine geoglyphs from the Peruvian Nazca Lines, two of which are included in this text. First, I present a theory on the figure carved into a mountainside at Nazca, commonly referred to as the "Astronaut." Secondly, I delve into the Monkey geoglyph, which has revealed extraordinary insights and messages.

These discoveries represent a fundamental shift in our approach to pyramid research, planetary systems, and cosmic understanding.

THE GIZA PARK PROTOCOL HOW TO START THE SUN

This "New Paradigm" is not merely about pyramids but encompasses a broader comprehension of our planet and the universe.

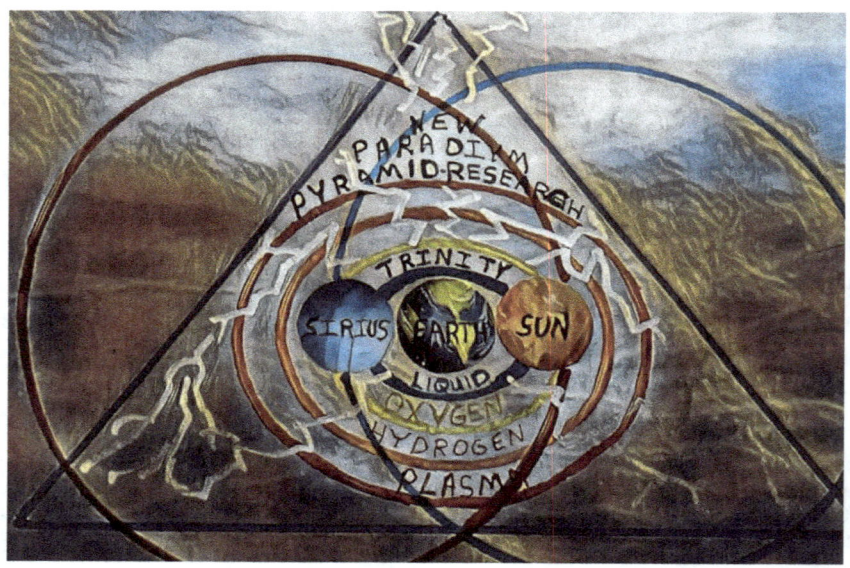

Once we lift our perspectives above the "Land of Oz" approach to pyramid research, we stand to gain more knowledge in twenty years than has been achieved over centuries. I recently came across a post stating that Egypt is introducing a Pyramid Research curriculum at a local university.

This suggests that Egypt may be ready to pursue the truth with integrity in pyramid systems research. However, this will require breaking free from decades of "Hollywood history," which has glamorised fairy tales and fabrications. Egypt has the greatest opportunity of any nation to unlock the secrets of pyramid systems. The complexity and scale of these natural systems, engineered with precision, seem to demand our attention. They metaphorically say: "Pay attention. Look over here."

If Egypt were to study these systems with accuracy and dedication, the knowledge gained could revolutionise human civilisation, contributing to the best and safest future for humanity. Sadly, much of Egypt's history has been reduced to entertainment and tourism, rather than a pursuit of truth or accurate historical understanding. Revealing precise history would challenge their existing religious narratives, and this hesitancy continues to obscure progress.

Looking back is essential for understanding ancient systems, and I often do so. At times, I feel a profound and almost indigenous connection with the Egyptian Thoth and the Greek Hermes. Reflecting on the depth of knowledge and expertise required by these ancient master builders, I wonder: what would we call them today in scientific terms?

When I began compiling the list of cosmic and Earth systems essential to the operation of these ancient pyramid systems, the sheer breadth was staggering. It involves studies of volcanic systems, astronomical alignments, tectonic plates, hydraulic continental divides, plasma, piezoelectricity, scalar energy, Earth's magnetic fields, significant mineral deposits and their polarities, gravity anomalies, and geoidal geography.

Additionally, ley line alignments, bioenergy grids, vortex systems, the Mediterranean and its surrounding seas, weather patterns, cyclones, hurricane formations, geological morphology, cosmic radio frequencies, reinterpretation of Egyptian hieroglyphs, the Nazca Lines, and Earth's atmospheric systems all play a role.

Adding all this together, Hermes and Thoth would likely be called "Cosmic Geomorphologists" today. Their expertise blankets multiple fields of science and introduces branches like "Archaeoastronomy" to enhance our understanding of environments and history. Isn't this precisely what we should strive for? To accomplish this with a spirit of joy and zest.

I find the title fitting. My forensic and energy experiences, along with my discoveries, have stemmed from a new way of examining what we've been falsely led to believe were tombs constructed by wandering desert tribes.

As part of this introduction, I invite you to explore some compelling quotes by Hermes Trismegistus and Edgar Cayce. These may resonate with you, the reader, as you make new discoveries across the globe.

The Emerald Tablets

By Hermes Trismegistus

With Translations Parallel to Pyramids

By Captain Leo Walton

Thousands of years ago, a priest named Thoth from the sinking land of Atlantis is said to have formed a colony in Egypt. He is credited with constructing the Great Pyramid of Giza. According to legend, Thoth carved texts onto emerald tablets, written in his native Atlantean language, and placed them in the King's Chamber of the Great Pyramid.

Many, especially the Greeks, believe Hermes and Thoth are one and the same. Their writings, including the *Corpus Hermeticum*, represent a collection of philosophical, religious, theological, and theosophical treatises. These works significantly influenced the development of alchemy and hermeticism, forming the foundation of the alchemist's philosophy of nature.

During my early research on the Great Pyramid, I began noticing a recurring natural relationship between the seven Hermetic Principles, which shaped my theories. Below are these principles as concisely stated as possible. After extensive study, I decided to use the original Latin version, which I believe most accurately reflects the essence of the Emerald Tablets.

The Seven Hermetic Principles

1) The Principle of Mentalism

"The All is mind; the Universe is mental."

2) The Principle of Correspondence

"As above, so below; as below, so above."

3) The Principle of Vibration

"Nothing rests; everything moves; everything vibrates."

4) The Principle of Polarity

"Everything is dual; everything has poles; everything has its pair of opposites. Like and unlike are the same; opposites are identical in nature but differ in degree. Extremes meet; all truths are but half-truths; all paradoxes may be reconciled."

5) The Principle of Rhythm

"Everything flows, in and out; everything has its tides; all things rise and fall. The pendulum swing manifests in everything; the swing to the right is the measure of the swing to the left. Rhythm compensates."

6) The Principle of Cause and Effect

"Every cause has its effect; every effect has its cause. Everything happens according to law; chance is but a name for law unrecognised. There are many planes of causation, but nothing escapes the law."

7) The Principle of Gender

"Gender is in everything; everything has its masculine and feminine principles. Gender manifests on all planes."

Let's begin:

1) Principle of Mentalism

"The All is mind; the Universe is Mental."

This first law is self-explanatory, so I will move directly to the discussion of:

2) Principle of Correspondence

"As above, so below; as below, so above."

If we focus on the Earth and compare it to the sky above—including the atmosphere, the Sun, the Moon, the planets, and all the celestial stars—we see a profound reflection. Below the Earth's surface lie the core, hydraulic continental divides, volcanic systems, rivers, and mineral deposits, both stone and metallic.

Thus, as the heavens above, so is the Earth below; and as below the Earth, so is above.

To illustrate this further using the Great Pyramid: As the Great Pyramid stands above, so are the Earth's systems reflected below; and as the Earth's systems lie below, so too does the Great Pyramid mirror them above.

3) Principle of Vibration

"Nothing rests; everything moves; everything vibrates."

The Earth rotates anti-clockwise at 1,000 mph. Its core, however, rotates in the opposite direction—clockwise—at a much slower and variable speed influenced by numerous factors. This dynamic interplay generates the Schumann Resonance Frequency, matching the human heart's vibration at 7.83 MHz.

The Great Pyramid moves at the same speed as the Earth and interacts with vibrations in three key ways:

Cadman's Pump: Positioned 100 feet below the pyramid, this pump operates with each stroke producing a pounding 3,200 PSI tide wave of water. This creates a resonance frequency vibration within the pyramid's interior.

Pyramid Shape: The structure itself captures the Schumann Resonance Frequency.

THE GIZA PARK PROTOCOL HOW TO START THE SUN

Internal Mechanics: The reaction chamber (commonly referred to as the King's Chamber) generates high-velocity combustion pressure. Like a boiling teapot, this pressure escapes through two 8-inch vents, producing a plaintive wail—akin to a howling wolf.

The pyramid's intricate design harnesses energy, forming a critical component of Egypt's weather and climate modification systems. This represents a fascinating example of quantum entanglement on Earth.

Due to the interconnectedness of the remaining principles, I will address them together:

4) Principle of Polarity

"Everything is dual; everything has poles; everything has its pair of opposites; like and unlike are the same; opposites are natural in nature but differ in degree; extremes meet; all truths are but half-truths; all paradoxes may be reconciled."

5) Principle of Rhythm

"Everything flows, out and in; everything has its tides; all things rise and fall; the pendulum swing manifests in everything; the swing to the right is the measure of the swing to the left; rhythm compensates."

6) Principle of Cause and Effect

"Every cause has its effect; every effect has its cause; everything happens according to law; chance is a name for a law not recognised; there are many planes of causation, but nothing escapes the law."

7) Principle of Gender

"Gender is in everything; everything has its masculine and feminine principles; gender manifests on all planes."

THE GIZA PARK PROTOCOL HOW TO START THE SUN

The Great Pyramid's geographical location on the Giza Plateau exemplifies these principles. Perfectly aligned to the cardinal points (north, south, east, west), its exterior acts as an antenna, capturing and manipulating masculine and feminine electromagnetic fields vertically. The pyramid's white limestone cladding reflects the Sun's positive solar energy, while its design protects the feminine Earth energies contained within.

Hermes stated:

"If we can shield an area of the Earth from the Sun's burning rays and build a structure capable of harnessing subtle and low-frequency energies, we can amplify natural strengths. Adding strength to strength ensures that these energies can conquer all interference and penetrate any field, from the most subtle air and water to the heaviest mass."

This design represents a form of alchemy, combining matter and energy into a marvellous system. Hermes described it as the method through which the world was created.

In his closing statement, Hermes declared:

"My name is Hermes Trismegistus, and this written communication is proof of the truth of my words. My understanding of the Principles of the Universe forms the basis of this Solar work project, which I now declare complete, tested, and operational."

These words left a profound impact on me. Comparing Giza with the *Emerald Tablets*, I identified over 100 quantum energy fields, most of which were harnessed within the limestone-insulated structure of the Great Pyramid.

Hermes' phrase, *"The combination of strengths, collected together, has the magnitude of power to conquer and penetrate all things,"* continues to resonate with me. It encapsulates the awe-inspiring capacity of the pyramid systems to interact with the most subtle energies or the densest masses on Earth.

THE GIZA PARK PROTOCOL HOW TO START THE SUN

Hermes makes some remarkable observations about the immense magnitude of the Great Pyramid of Giza's combined bio-energies and how they are maximised by their interaction. For me, this represents Quantum Entanglement at its most profound.

One of his most significant statements, taken from the *Emerald Tablets*, ties perfectly to his earlier comment. Known as the Principle of Correspondence, Hermes states: *"As above, so below; as below, so above."* This concept, attributed to Thoth/Hermes, forms the foundation and framework of all my findings.

But what does this mean precisely? As we progress on this journey, I believe the time will come to explore this concept in detail, especially when we have fully grasped the genius and cosmic knowledge required to design, construct, and determine the geographical location of this powerful system. The system utilises indigenous materials and harnesses available energies.

To date, I have written thousands of pages of hypothetical research theories and created several hundred detailed system drawings of the components associated with the Giza Plateau. The Principle of Correspondence, *"As above, so below,"* underpins my findings. These efforts have allowed me to present the most comprehensive hypothesis of the systems on the Giza Plateau and the broader "Old Kingdom" era since their inception.

My research has incorporated a wealth of ancient symbolic messaging and the study of mythological figures, particularly the Egyptian god Thoth and his Greek counterpart Hermes. I have unified these two powerful deities from different cultures to clarify their similarities. The Greeks' admiration and respect for Egyptian religious systems is evident in their adaptation of the Egyptian pantheon to suit their own cultural context. This blending demonstrates their reverence for Egyptian philosophy and their desire to integrate its fundamental concepts.

THE GIZA PARK PROTOCOL HOW TO START THE SUN

After immersing myself in the contributions of researchers, archaeologists, scientists, and authors, and studying numerous academic works, one individual stands out for his unique approach to Egyptian history: the famous psychic clairvoyant Edgar Cayce. Cayce amassed a vast collection of documented psychic readings, which he claimed were drawn from the Akashic Records. His work attracted influential figures, including the great inventor Nikola Tesla, who was a close friend and collaborator for over 40 years.

Cayce's non-profit organisation, the Association for Research and Enlightenment (ARE), was founded in 1931 in Virginia Beach, Virginia. Initially established as a hospital and research centre, it has become one of the largest private investors in research and archaeology on the Giza Plateau.

Virginia Beach remains the organisation's headquarters and houses the Atlantic University campus, which includes one of the world's largest libraries. The library contains over 80,000 volumes specialising in metaphysics, parapsychology, comparative religious studies, holistic health, and ancient civilisations. At its core are over 14,000 of Cayce's psychic readings, including more than 8,000 personal readings, preserved as a living memorial to his work. Today, ARE has centres in 31 countries.

I first visited the Virginia Beach centre in 1976 with my sister, who was an admirer of Cayce. Although I was unfamiliar with him at the time, the visit sparked my interest. Years later, I had the extraordinary experience of dining twice at the house where Cayce lived before his death. I often wonder if Tesla himself may have visited this residence.

Interestingly, Tesla and Einstein contributed to the Philadelphia Experiment, conducted just a mile from this location—a fascinating and eerie coincidence. I feel fortunate to have been so close to this institution and to have accessed its unparalleled collection of knowledge, much of which is unavailable elsewhere.

THE GIZA PARK PROTOCOL HOW TO START THE SUN

In 2014, I was invited by the International Association of Near-Death Studies to lecture on my near-death experiences and to introduce my theories on the Giza Plateau Pyramid systems.

Cayce provided over 1,200 psychic readings focused on Egypt, offering insights into the origins, purpose, and prophecies of the Great Pyramid and the Sphinx, as well as his own past-life experiences as the priest Ra-Ta. His readings explored spiritual unification and the ancient, now-lost civilisation of Atlantis, referenced by Plato in his dialogues *Timaeus* and *Critias* in the 5th century BCE.

It is in this context that I feel compelled to share some of Cayce's visions. These glimpses provide a broader understanding of the Giza Plateau, which I have dubbed the "Giza Industrial Park," as well as Atlantis and other fascinating aspects related to my discoveries.

Edgar believed that the history of humankind spanned more than 10 million years. He utilised his Trance State method of "Retrocognition"—a technique he called it—that repeatedly provided uncanny and highly accurate information. His knowledge of a Jewish sect known as the Essenes, the group to which Jesus belonged, was shared by Edgar 11 years before the Dead Sea Scrolls were discovered in 1947, two years after his death. It was only after years of archaeological excavations at these sacred sites that Cayce's psychic information was verified as accurate.

EC Reading 364-13

Edgar claimed that the Nile River once flowed into the Atlantic Ocean. What is now the Sahara Desert was once a fertile, inhabited land. The central part of the United States, specifically the Mississippi Basin, was entirely submerged under the ocean, with only the plateau remaining above water. The regions now comprising Nevada, Utah, and Arizona formed a significant part of what we now know as the United States.

Interestingly, Science Magazine confirmed this Cayce theory in 1986, based on NASA radar imagery captured aboard the Space Shuttle. These satellite images, coupled with later on-site archaeological investigations, provided evidence that the Nile River had indeed changed its course, once flowing across the Sahara and into the Atlantic Ocean.

EC Reading 440 – Atlantean Power Plants

Edgar describes the materials and functions used to construct the Atlantean Pyramid. He said:

"About the Firestone, the entity's activities then made such applications as dealt with within the constructive and destructive forces during that period."

"It would be well that there be given something of a description of this so that it may be understood better by the entity present."

"In the centre of the building, which today could be said to be lined with a non-conductive stone, something akin to asbestos, along with other non-conductors now being manufactured in England, under a name that is well known to many who deal in such things."

"The structure above the stone was an oval or dome, where there could be a portion for rolling back, allowing the activity of the stars and the concentration of energies emanating from bodies on fire, as well as elements not found in the Earth's atmosphere, to interact."

"The concentration through the prisms of glass, as we would call it in the present, acted in such a way that it influenced the instruments connected to various modes of travel through induction methods, producing effects akin to modern-day remote control via radio vibrations or directional forces. The power from the stone acted upon the motivating forces in the crafts themselves."

Atlantean Hall of Records

THE GIZA PARK PROTOCOL HOW TO START THE SUN

Edgar Cayce's predictions and comments about the Atlantean Hall of Records, which he claimed was located beneath the Sphinx, have generated much excitement and controversy.

Studying Edgar Cayce's language style has revealed that his comments can sometimes be ambiguous, allowing for numerous translation interpretations. My own discoveries have taught me valuable lessons in interpretation, translation, and symbolic representation. Consequently, I am comfortable with Cayce's prose and have found his insights enlightening on many occasions.

As we progress on this journey, I encourage you to reflect on Edgar's and Hermes' teachings and effortlessly integrate them into your own perceptions.

EC 378-16, 10/29/33

(Q) What does the sealed room of the Atlantean Records contain?

Edgar's answer:

"This position lies, as the Sun rises from the waters, and the line of the shadow falls between the paws of the Sphinx, which was later set as the sentinel or guard, and which may not be entered from the connecting chambers of the Sphinx's right paw until the time has been fulfilled, when the changes must be active in this sphere of man's experience."

"Between, then, the Sphinx and the river."

EC 3916-15, 1/19/34

(Q) Who will uncover the history of the past in record form, said to be near the Sphinx in Egypt?

Edgar's answer:

"As was set in those records of the Law of One in Atlantis, that there would come three who would make of the perfect way of life...

These, then, make themselves that channel. For, as He has given, it is not mine to declare who will proclaim, but they have made of themselves such a measure of their experiences as worthy of proclamation."

EC 440-5, 12/20/33

"The records of the methods of construction of the Atlantean Firestone are located in three places on Earth as it stands today: in the sunken portions of Atlantis or Poseidon, near Bimini, and in the temple records in Egypt. The records were also carried to what is now Yucatán in America."

I relate to several of these quotes, and it is for this reason that I am sharing them with you, to assist you in your journey.

Chapter Two:
Biography

At first glance, this volume may appear to be a collection of unrelated, eclectic topics with little connection to one another.

My aim, however, is to propose a new model for structuring pyramid systems research. I seek to establish an entirely new paradigm regarding how and why the complex natural systems of our environment—and beyond—are directly connected to these ancient pyramid systems.

By the time you reach the end of this treatise, I hope you will have gained an understanding of how these diverse, dedicated components function as part of an overarching industrial system.

By identifying each piece of this marvellous puzzle—left for us to learn from and gain wisdom—we can develop a fresh perspective on how these components fit into the integrated system as a whole.

If the great inventor and physicist Nikola Tesla were alive today, I believe he would have appreciated the theme of this book. This volume, therefore, explores the newly identified facets necessary for understanding the complexities of ancient pyramid systems.

It encompasses a wide-ranging collection of disciplines, including physics, chemistry, biology, atmospheric science, astronomy, geology, volcanology, hydrology, and cosmic phenomena, all required to extract the knowledge these systems hold.

In this book, I reference Tesla's famous quote: "If you want to find the secrets of the universe, think in terms of energy, frequency, and vibration."

THE GIZA PARK PROTOCOL HOW TO START THE SUN

As a lifelong investigator, I am deeply passionate about uncovering ancient secrets and exploring the invisible world of bioenergetics and the indigenous and cosmically quantum entanglement systems.

I aim to guide you on a journey, providing you with "charts of navigation" to inspire discovery and awareness. My hope is to help you rediscover a lost connection with our natural and indigenous links to the universe, using Tesla's genius as a foundation.

The information in this text is based on research conducted over the course of more than fifteen years. During this time, I became profoundly fascinated by and dedicated to studying the Great Pyramid of Giza in Egypt.

THE GIZA PARK PROTOCOL HOW TO START THE SUN

When I say I was focused, I should clarify—I was obsessed. I often joke that I became the "Mashed Potato Guy" from the 1977 Spielberg classic *Close Encounters of the Third Kind*. My research began solely with the Great Pyramid of Giza, the last remaining physical structure of the original Seven Wonders of the Ancient World.

I discovered that most of the research conducted so far had centred on how the original builders—whoever they were—managed to accomplish such a momentous task with astonishing precision, a feat we are still unable to replicate today.

THE GIZA PARK PROTOCOL HOW TO START THE SUN

My focus gradually expanded in scope. Eventually, I decided I no longer cared how the Great Pyramid was built; I couldn't see any significant benefit in knowing how. Instead, my questions began to centre around why the Great Pyramid was constructed.

It became increasingly clear that the Great Pyramid was never intended to serve as a tomb, despite decades of claims to the contrary by Egyptian archaeologists. When you consider the monumental task of constructing the entire Giza Plateau, other pressing questions emerge: who would invest such vast resources and time into this endeavour, and what benefit did the completed project provide them?

THE GIZA PARK PROTOCOL HOW TO START THE SUN

Why was it built in the middle of the desert, far from any known civilisation? The logistics of this project are staggering, seemingly impossible to bring to fruition by any conventional means.

Fortunately, the discovery of significant works by an extensive group of "pyramid-obsessed" researchers over the decades has provided valuable insights. Visionaries such as Kunkel, Parr, Petrie, DeSalvo, Meyer, Mehler, and, more recently, John Cadman and Christopher Dunn, have contributed immensely to this field of study.

Reading Christopher Dunn's eloquently phrased work, *The Giza Power Plant: Technologies of Ancient Egypt*, steered me in a more rational and familiar direction—exploring the Great Pyramid from a mechanical perspective. Given my extensive professional experience in mechanical systems spanning decades, I immediately connected with Dunn's mindset.

THE GIZA PARK PROTOCOL HOW TO START THE SUN

Once I embraced the idea that the Great Pyramid was most likely a mechanical device, everything began to fall into place. It became far easier to comprehend that the Great Pyramid was not an isolated structure but rather a single mechanical component within a vast and extraordinarily complex network of ancient industrial systems.

The further I explored and dug, the more often I was surprised with my discoveries. One after another, component after component, they are linked in a seemingly never-ending parts inventory, like a box of children's Legos. Over time and 1000s hours invested in investigative research, I eventually determined that all structures built on the Giza Plateau and below the ground were an integral part of this one dedicated system.

THE GIZA PARK PROTOCOL HOW TO START THE SUN

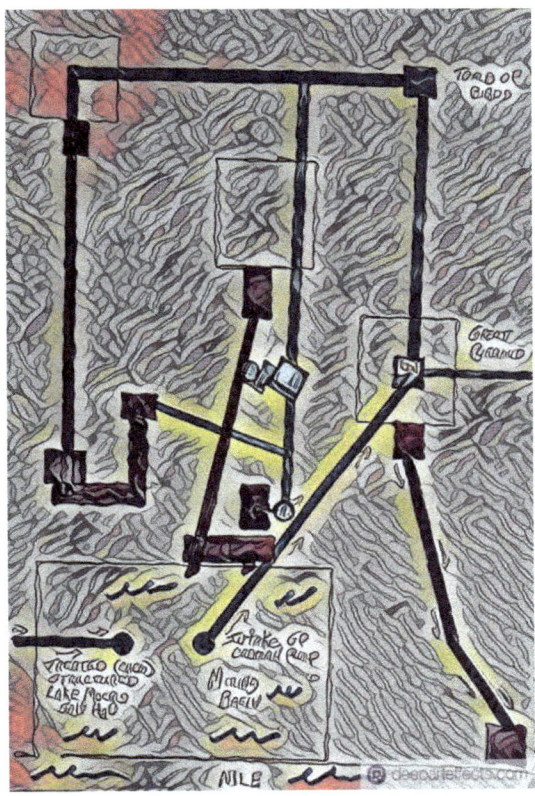

As my research unfolded, it felt like assembling a puzzle, piece by piece. I discovered that the Giza Plateau was not an isolated system but part of an interconnected network of industrially designed systems. This complex extended beyond Egypt, incorporating other systems from the Old Kingdom period and far-reaching locations around the world.

My geographical research expanded to include pyramid systems as far away as the Teotihuacan pyramids in Mexico. I recently recognised the Bosnian Pyramid systems as a vital component of ancient weather and climate control modification systems. Adding the Chinese pyramids and the Angkor Wat systems to my list brought the total to five ancient pyramid systems—each an interactive system functioning simultaneously. This led me to ask the ultimate question: *Why?*

THE GIZA PARK PROTOCOL HOW TO START THE SUN

Using my forensic investigative methods and mechanical engineering skills, I quickly realised that this desert-based industrial complex served multiple purposes, far beyond what was initially apparent. By fragmenting and analysing the various systems, I gathered sufficient evidence to apply reverse engineering techniques, which formed the basis of the theories you are now reading.

I began to view the Great Pyramid as a "Swiss Army Knife," a device of incredible versatility and capability, designed with a multitude of functions in mind. My decades of training and experience in mechanical engineering, research and development, and large-scale industrial projects provided the foundation for these insights. My background includes nuclear-marine shipbuilding, two patents, corporate management training across multiple systems (American, German, British, and Japanese), as well as roles as a USCG Merchant Marine Captain, a founding member of US Homeland Security, a successful small business owner, and a marine forensics investigator and expert courtroom witness.

I used my extensive expertise—gained through over 2,500 forensic investigations—and combined it with the methods I created to lay the groundwork for my ancient hypothetical theories. My investigative approach has always involved collecting and meticulously documenting evidence from every conceivable source, leaving no stone unturned. Like breadcrumbs scattered across the globe, these fragments of corroborating evidence are crucial. I believe humanity must act as a collective community to uncover the lost truths of our history, planet, and cosmos.

I firmly believe that much of what we have been taught about our historical past, Earth, and the cosmos has been deliberately corrupted and fabricated, designed to keep the masses in ignorance and serve an agenda of profit and control.

My research has also involved studying ancient symbolic messaging and mythological figures, particularly the Egyptian gods Thoth and Hermes, which the Greeks adopted and adapted into their own pantheon. To avoid confusion, I treat these two deities as one entity, recognising the Greeks' admiration for Egyptian philosophy and their efforts to integrate Egyptian concepts into their cultural framework.

After digesting countless academic studies, theses, and significant contributions from researchers, investigators, archaeologists, scientists, and authors, one individual stands out: Edgar Cayce. Known as the "Sleeping Prophet," Cayce approached Egyptian history in a unique and unconventional way. As one of the largest private investors in Giza Plateau research, his insights—though controversial—add a fascinating dimension to the search for answers.

I believe including some of Cayce's perspectives enriches the broader narrative. After all, if I were assembling a team to explore the mysteries of the Great Pyramid, I'd want a credible psychic on board too.

Chapter Three:
Earth & Egyptian History

One exciting connection I will explore as we progress is that many scientists believe the ancient Egyptian and Mayan civilisations each lasted nearly 3,000 years. My research into both cultures continues to uncover similarities between these two civilisations. Interestingly, scientists assert that the two periods align chronologically.

While I aim to keep the timeframes of these civilisations simple, complications arise when dating events accurately. This becomes especially problematic in light of the false narrative that human existence spans only 6,000 years, bolstered by fabricated evidence to support a fictitious history.

The Egyptians unified their nation around 3100 BC, and Alexander the Great conquered Egypt in 332 BC, ending the Pharaonic dynasties and ushering in the Ptolemaic period. Furthermore, Egypt experienced a Dry Period beginning around 3000 BC, characterised by inconsistent rainfall that led to devastating droughts, famine, and disease by 2500 BC. This weakened the nation and made it a target for invasion.

A succession of foreign powers conquered Egypt, plundering its wealth and treasures. The Hyksos, Libyans, Nubians, Assyrians, and Achaemenid Persians were among the attackers. Eventually, Alexander the Great, one of history's most brilliant military minds, entered Egypt during its period of vulnerability. He effectively ended the remnants of Egyptian authority, establishing Alexandria in his name.

Alexandria became the seat of the Macedonian Greek royal dynasty, known as the Ptolemaic Kingdom, which was founded in 305 BC under Ptolemy I Soter, a trusted general of Alexander. This kingdom persisted for nearly 300 years until the death of Cleopatra VII in 30 BC, following the Roman invasion.

On the other side of the globe, the Mayan civilisation is believed to have emerged in Mesoamerica around 2500 BC, coinciding with the Egyptian droughts. This marks the beginning of the Pre-Classic period, which lasted until approximately 250 AD.

The parallels between the ancient Egyptians and the Mayans suggest significant synchronistic possibilities. The Dry Period in Egypt, which peaked around 2500 BC, was catastrophic, leading to widespread crop failures, starvation, and internal conflicts. By 2200 BC, Egypt's Old Kingdom had collapsed, with civil wars, invasions, and looting contributing to its downfall.

Historians remain puzzled by the fate of the ancient Egyptians, as large populations seemingly vanished without trace. Similarly, the origins of their advanced academic knowledge and construction techniques remain shrouded in mystery. Some evidence suggests that Egyptian seafarers may have ventured far afield, possibly even reaching the Americas. Reports of Egyptian artefacts found in caves south of Chicago and hieroglyphs discovered in the Grand Canyon have fuelled speculation, though these sites remain closed to the public under government protection.

Could there have been a great exodus from Egypt during this catastrophic period? Unfortunately, little evidence survives to provide an accurate timeline. Instead, much of what we rely on is drawn from religious texts or speculative historical reconstructions.

Two significant biblical events are worth mentioning here: the Great Flood and the Exodus. The Bible claims the flood occurred before 3000 BC, while the Exodus is dated to 2666 BC. These events align broadly with the scientifically documented Dry Period in Egypt, which spanned from 3000 BC to 2500 BC.

THE GIZA PARK PROTOCOL HOW TO START THE SUN

My years of archaeological research have taught me that providing precise timelines for undocumented historical events is a challenging task. Advances in scientific methods and technology continually refine our understanding of ancient history, yet uncertainties remain.

Looking further back, the Bonneville Flood in North America occurred around 14,500 BC, coinciding with the end of the last Ice Age. By 9,000 BC, sheep herders had begun migrating to the Mediterranean Fertile Crescent, as melting ice made the region habitable. Evidence suggests that agriculture in Egypt began near 7700 BC, with wheat as the first crop, followed by peas around 6500 BC.

In upcoming chapters, I will reference the Summer Solstice and the Lions Gate Portal of 7000 BC as a framework for discussing key astronomical and geographical alignments. This timeline provides a structured way to examine pyramid systems and their significance.

Imagine standing before the Great Sphinx of Giza during the Summer Solstice of 7000 BC, gazing eastward as the heliacal rise of Sirius marks the dawn. The planetary alignments at this moment would have held immense significance, influencing the design and purpose of the pyramid systems.

This visualisation allows us to reconstruct and explore an era that continues to captivate and mystify. By examining these ancient civilisations through a scientific lens, we can uncover deeper truths about their legacy and connections.

THE GIZA PARK PROTOCOL HOW TO START THE SUN

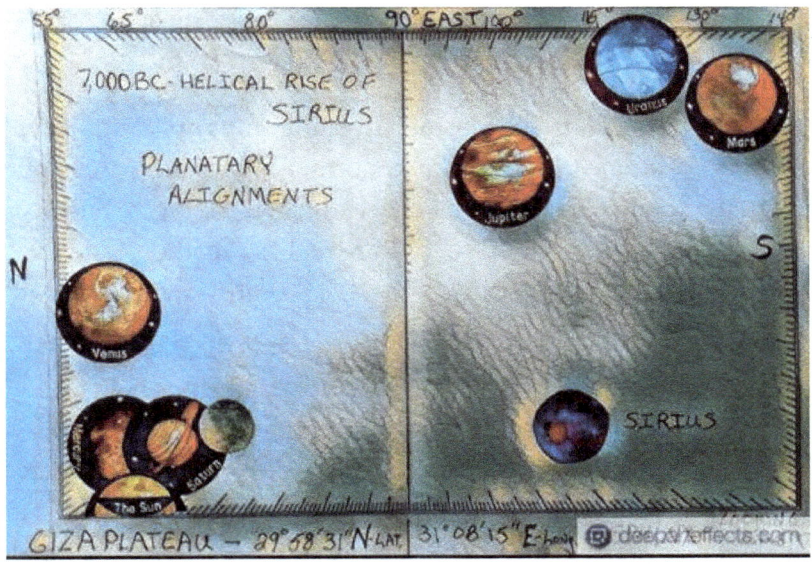

This image exemplifies events that occurred at the Giza Plateau during the heliacal rise of Sirius.

This celestial event, marking the heliacal sunrise of the two bi-solar suns—masculine and feminine—was also associated by the Egyptians with the onset of the annual rainy season and the nutrient-rich soil brought by the Nile floods.

I have grown accustomed to using this image as a central reference throughout this treatise. It serves as a baseline to arrange events chronologically and help readers track developments linearly. Given the breadth of topics covered, I decided it was best to designate a single reference point to anchor the discussion.

Once you have absorbed my theories, one thing will become clear: planetary alignments and recurring cyclical events within our solar system have been instrumental in creating functional pyramid systems on Earth, benefiting humanity.

In this analysis, I have considered the 8,000 BC timeframe suggested by scientists as a possible starting point for the rainy season. However, we must acknowledge that this estimate could be off by 1,000 years or more.

To visually align the cosmic dynamics involved in Egypt's weather and climate modification systems, I rely on a single cyclical event as a reference point. Using all available data, I have developed a step-by-step, organised protocol to reveal the methods and cycles behind these systems, creating a cohesive narrative in a world where none has previously existed.

The termination of the Ice Age had profound effects on humanity, making migration easier and safer. With an improved climate, abundant vegetation, and accessible water resources, humankind seized the opportunity to establish new cultures and civilisations. By 2500 BC, this unprecedented development had led to the settlement of every continent except Antarctica due to its harsh environmental conditions.

Now, let me outline some key timeframes during which new civilisations emerged, highlighting significant events:

The Great Exodus of Egyptians was not confined to a single event or civilisation but was part of a broader, global phenomenon. Among these events is the controversial case of the 300 Egyptian hieroglyphs carved in stone in Australia. Dr R.M. de Jonge, who investigated and translated the Kariong Hieroglyphs—also referred to as the Gosford Glyphs—claimed the site was an Egyptian monument. According to his research, it commemorates an expedition led by Pharaoh Khufu's son, Lord Nefer-Ti-Ru. As the story goes, Lord Nefer-Ti-Ru died after being bitten by a poisonous Australian snake.

These hieroglyphs, depicting boats, chickens, dogs, owls, stick figures, and two cartouches (one of Pharaoh Khufu, the second king of the Fourth Dynasty, 2637–2614 BC, and an unidentified figure), support my theory that the Egyptians explored global sailing routes.

THE GIZA PARK PROTOCOL HOW TO START THE SUN

Earlier seafaring civilisations, particularly the early Greeks, were also skilled ocean sailors who likely voyaged before the Egyptians. I will present further data to support these claims.

6,000 to 2,900 BC: The Neolithic period saw early Greek populations voyaging the Mediterranean islands.

3,200 to 1,100 BC: Cycladic civilisation flourished, with early Greek fishermen inhabiting the Cyclades islands, including the prominent island of Delos.

3,000 to 1,200 BC: The Bronze Age in Greece began. In 1990, sunken Greek ships discovered off the coast of Turkey revealed tonnes of copper and zinc used to create bronze (circa 1300 BC). Metallurgical analysis indicated that the copper originated from Cyprus, while the tin likely came from Afghanistan, Lebanon, or Israel.

2,700 to 1,500 BC: The Minoan civilisation emerged as the first major Greek civilisation, located on the island of Crete. Crete became a hub for trading natural resources such as timber, wool, tin, copper, bronze, and ceramics. The pottery wheel was also invented here, revolutionising the ceramics industry.

2,637 to 2,614 BC: This narrow 23-year window aligns with the timeframe during which the Egyptian hieroglyphs were discovered in northern Australia.

2,500 BC: Excavations and construction began at the Hypogeum of Hal-Saflieni in Paola, Malta. The Indus Valley civilisation emerged on the Indian subcontinent.

2,400 BC: Construction of Stonehenge began.

2,300 BC: The Unetice culture emerged in modern-day Czechia.

2,234 BC: Sargon the Great established the Akkadian Empire, conquering all of Sumer.

THE GIZA PARK PROTOCOL HOW TO START THE SUN

2,100 BC: The rule of Ur-Namma marked the beginning of a new dynasty. The Xia and Shang dynasties began in Ancient China.

2,000 BC: The Mayan civilisation emerged in Mesoamerica. Early Nubian culture developed. The Aegean civilisation began.

1,900 to 1,100 BC: The Mycenaean civilisation flourished in continental Greece, centred around Mycenae. According to myth, Agamemnon ruled here, as documented in Homer's *Iliad* and *Odyssey*. The Mycenaean civilisation also developed Linear B, the earliest known written Greek language.

1792 BC, Age of Aries: King Hammurabi established the capital of Babylonia.

1600 BC, Age of Aries: The Kurgan culture began.

1600 BC, Age of Aries: The Mycenaean culture emerged.

1500 BC, Age of Aries: The Aryan invasion occurred.

1200 BC, Age of Aries: The Pueblo Indians and Olmec Indians began developing their cultures.

1000 BC, Age of Aries: Celtic culture emerged.

753 BC, Age of Aries: Rome was founded.

600 BC, Age of Aries: The Olmec writing system was created.

200 BC, Age of Aries: The Mayan script was developed.

The information on volcanic eruptions shared in this text is primarily based on two significant research projects.

The first is the U.S. National Science Foundation's Polar Program in Antarctica, which received additional contributions from Belgium, Canada, China, Denmark, France, Germany, Iceland, Japan, Korea, The Netherlands, Sweden, Switzerland, and the United Kingdom.

The second is the North Greenland Eemian Ice Drilling Project (NEEM), led by the Niels Bohr Institute and supported by 14 additional countries.

These research teams are conducting two parallel projects to extract ice core samples from Greenland in the Northern Hemisphere and Antarctica in the Southern Hemisphere. By comparing these ice core samples, scientists can identify similarities and differences in climatic and volcanic activity across the two hemispheres.

For the first time, I am providing a list that identifies ancient periods of significant events. I have chosen 14,000 BC, the Age of Libra, as the starting point for my hypotheses and theories. To strengthen the foundation of my research, I have included scientific information on volcanic activity, historically documented from 10,900 BC to modern times, revealing numerous catastrophic events throughout history. These events help fill in the gaps of my customised version of the chart of "The Great Year."

THE GIZA PARK PROTOCOL HOW TO START THE SUN

Note: Scientists generally agree that Mt. Etna in Italy is the oldest volcano on Earth, estimated to be between 350,000 and 500,000 years old. Additionally, most of the 1,500 active volcanoes today are presumed to be less than 100,000 years old.

Between 73,000 BC and 10,900 BC, scientists have identified at least 75 known volcanic eruptions worldwide. However, due to the absence of systematic historical record-keeping before 5000 BC, the data on these eruptions is insufficient for precise dating and damage assessment.

On a positive note, modern scientific advancements, particularly research by Cole-Dai and his colleagues, have significantly improved our understanding of volcanic activity. Their work on the West Antarctica Ice Sheet Divide involved drilling deep into the ice core and extracting samples. These samples provided new physical evidence of explosive volcanic eruptions in the Southern Hemisphere during the Holocene period, spanning the last 11,000 years.

Using high-resolution measuring equipment, researchers analysed the chemical composition of the 3,400-metre-long ice core. This enabled them to determine the chronological age of 426 significant eruptions since 9000 BC. They pinpointed the precise years when these eruptions occurred, offering unprecedented accuracy.

The atmospheric aerosol mass loading of climate-impacting sulfur, as measured from volcanic sulfite deposition, revealed an extraordinary period in the second half of the 17th century BC (1700–1600 BC). During this time, the Thera volcano in the eastern Mediterranean is believed to have erupted. Notably, volcanic signals from this period are synchronised with three eruptions detected in Greenland Ice Cores. These findings strongly suggest that the Thera eruption, along with other eruptions, occurred in low latitudes and had a global impact.

THE GIZA PARK PROTOCOL HOW TO START THE SUN

I have utilised this data to identify significant volcanic eruptions and other events on my chart. Each event corresponds to an astrological age, with each age lasting approximately 2,000 years.

For example, 12,900 BC falls 1,100 years into the Age of Libra, which lasted from 14,000 BC to 12,000 BC. The Age of Libra is renowned for its focus on symmetry, balance, and equilibrium in all aspects of life.

During this period, the Weather and Climate Modification Systems operated in two stages: cycling off and then cycling on again 1,200 years later to begin the second phase.

Our two suns—Sirius and our Sun—share an external orbital axis. Entering the Age of Libra in 14,000 BC, Sirius and our Sun formed the primary components of our cyclic binary satellite system. Scientific exploration suggests that physical evidence supports the claim that the last Ice Age began to melt 500 years before the onset of the Age of Libra, around 14,500 BC.

The year 14,500 BC is particularly significant, as it marks one of the most catastrophic events in North America: the Bonneville Flood. This event remains a mystery to scientists, who continue to debate its causes and consequences. Its impact on the landscape of the Pacific Northwest was dramatic, contributing to the formation of the Great Basin, which spans several states in the United States.

The **Bonneville flood** was a catastrophic flooding event in the last ice age, which involved massive amounts of water inundating parts of southern Idaho and eastern Washington along the course of the Snake River. Unlike the Missoula Floods, which also occurred during the same period in the Pacific Northwest, the Bonneville flood happened only once. It is believed to be the second-largest flood in known geologic history.[1][clarification needed]

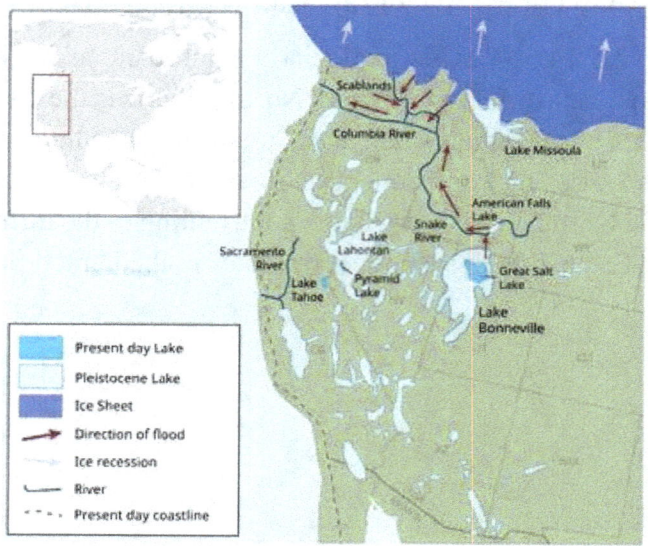

Map of Pleistocene lakes in the Western US, showing the path of the Bonneville Flood along the Snake River

THE GIZA PARK PROTOCOL HOW TO START THE SUN

The biggest mystery about flooding events is that they are typically recurring phenomena, not isolated, earth-shaping events like the Bonneville Flood. Further scientific research has uncovered evidence of sunken landmasses and measurements indicating a 30-foot rise in ocean levels around 11,500 BC, during the Age of Virgo.

Modern research supports a 3,000-year timeframe during which melted ice from frozen latitudes flowed into the oceans, causing a significant global rise in sea levels. This data provides a valuable connection, and I will explain its significance. While historical records are limited, scientific claims indicate that around 14,500 BC, vast quantities of water began moving from previously ice-covered regions.

By comparing these dates—14,500 BC to 11,500 BC—we arrive at a 3,000-year period. During this time, sea levels rose by approximately 30 feet globally. This translates to a rise of 10 feet per 1,000 years.

This straightforward equation—30 feet over 3,000 years—will be useful for future determinations of timeframes. I always rely on credible scientific sources to accurately establish logical timelines.

I want to highlight a groundbreaking discovery from recent radiocarbon dating conducted on organic samples found at the Bosnian Pyramid of the Sun. During his archaeological investigations, Dr. Semir Osmanagić unearthed organic material—leaves and sticks—trapped between the layers of artificial concrete cladding on the Pyramid's perfectly aligned and sloping sides. Radiocarbon dating confirmed that this material is 32,000 years old.

This is the first instance of scientifically tested and verified data proving that the Bosnian Pyramid of the Sun was constructed at least 32,000 years ago. The organic material could not have been deposited naturally beneath the shingled layers of artificial concrete.

THE GIZA PARK PROTOCOL HOW TO START THE SUN

As we delve deeper into these findings, I will return to a more detailed discussion of this 32,000 BC period to develop hypotheses that offer logical explanations for the events of that time.

Around 14,500 BC, during the Age of Scorpio, the Bonneville Flood occurred. This event is considered a direct result of the Ice Age melt, and scientists have used it as a reference point to pinpoint the beginning of the Ice Age's decline.

It's important to note that the Bonneville Flood coincided with the first phase of the Weather and Climate Modification Systems, which began operations around 32,000 BC.

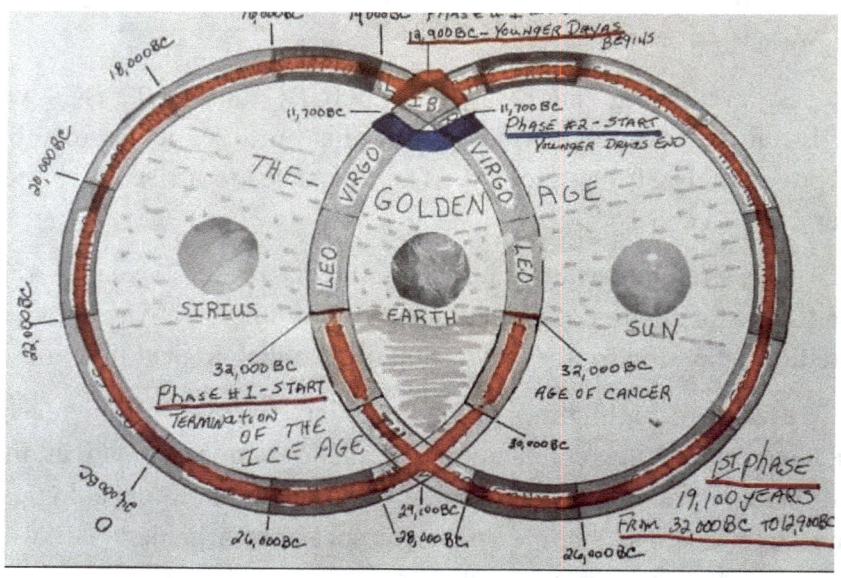

Following 19,100 years of cyclic Weather and Climate modification systems operations, that cycle ended in 12,900 BC during the Age of Libra, marking the beginning of the 1,200-year Younger Dryas Period.

12,900 BC Termination of the Ice Age 1st Phase agenda complete, project closed.

THE GIZA PARK PROTOCOL HOW TO START THE SUN

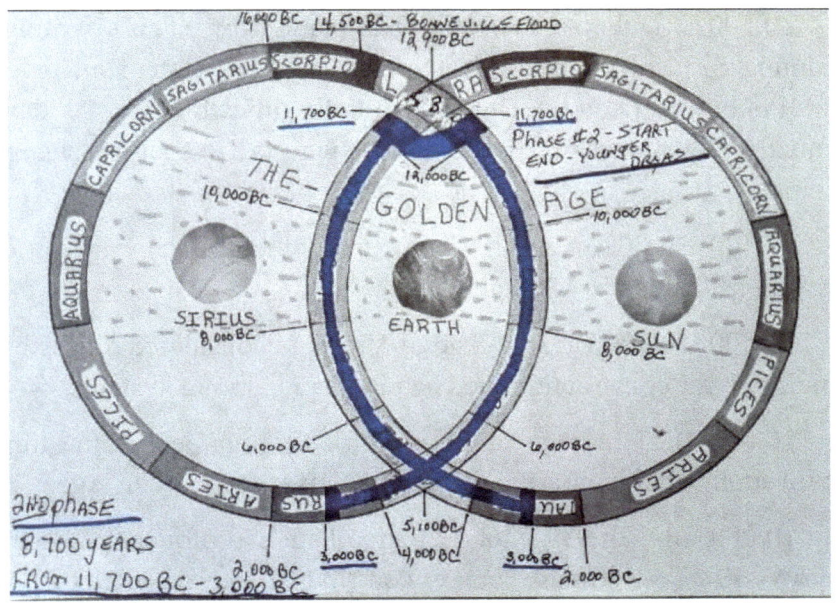

12,900 BC, the Age of Libra, marked the beginning of the Younger Dryas Period.

11,700 BC, the Age of Libra, saw the second phase of the Weather and Climate Modification Systems. The Younger Dryas Period, lasting 1,200 years, came to an end.

In 11,500 BC, the Age of Virgo, scientists confirmed research concluding that sea levels rose 30 feet in the preceding 3,000 years.

11,000 BC, the Age of Virgo, witnessed the eruption of Glacier Peak in Washington State, USA.

11,000 BC, the Age of Virgo, is the oldest date that the Ice Core Research project could accurately identify.

In 11,500 BC, the Age of Virgo, modern scientists determined the existence of sunken continents and a 160-foot rise in sea levels. Physical evidence indicates that ice was melting during the Ice Age.

10,900 BC, the Age of Virgo, saw the Lacher See eruption in the Eifel Mountains range of Rhineland, Germany.

THE GIZA PARK PROTOCOL HOW TO START THE SUN

10,500 BC, the Age of Virgo, is marked by Plato's writings claiming Sirius wasn't visible then—an unusual statement, given that Plato understood Sirius is naturally invisible for 70 days annually, a cyclic event. His comment suggests a deeper message to decipher.

10,500 BC, the Age of Virgo, also witnessed the eruption of Nevado de Toluca in Mexico.

10,500–10,000 BC, the Age of Virgo, is considered a possible timeframe for constructing the Giza Plateau pyramid systems.

10,000 BC, the Age of Leo, is widely regarded as the time civilisation began on Earth, also referred to as the Golden Age.

10,000 BC, the Age of Leo, marked a profound planetary energy shift. Two primary sources contributed: natural energy fields and contrived mechanisms using those natural fields.

The correlation between Sirius and the Sun was significant, as their orbits centred in Vesica Pisces formations. These events occurred on 3rd January and 4th July in what is now America. Additionally, the initiation of the Giza Plateau pyramid systems enhanced the Earth's incidental radiant energy capabilities.

9,460 BC, the Age of Leo, witnessed the Taupo Caldera eruption in New Zealand.

9,450 BC, the Age of Leo, marked the eruption of Mount Tongariro in New Zealand.

9,000 BC, the Age of Leo, saw the ice covering the mountains of the Mediterranean Crescent melt sufficiently, prompting shepherds to move their flocks to higher altitudes. This marked the first documented human activity in the region, previously covered with thick ice. Scientists also assert that the Sahara Desert became lush with vegetation, with rainfall filling empty caverns to form lakes.

THE GIZA PARK PROTOCOL HOW TO START THE SUN

8,230 BC, the Age of Leo, witnessed the Grimsvotn eruption in Iceland.

8,130 BC, the Age of Leo, marked another Taupo Caldera eruption in New Zealand.

8,000 BC, the Age of Cancer, saw a great flood that lasted for an extended period.

8,000 BC, the Age of Cancer, also marked the cultivation of potatoes and squash in Mesoamerica.

7,700 BC, the Age of Cancer, saw the cultivation of wheat as the first crop in Egypt.

7,560 BC, the Age of Cancer, marked the eruption of the Rotoma Caldera in New Zealand.

7,480 BC, the Age of Cancer, witnessed the eruption of Lvinaya Past in Iturup, Kuril Islands, Russia.

7,460 BC, the Age of Cancer, saw an eruption at Mount Pinatubo on Luzon Island, Philippines.

7,420 BC, the Age of Cancer, witnessed the eruption of Fisher Caldera on Unimak Island in the Aleutians.

7,000 BC, the Age of Cancer, saw the continuation of the great deluge, creating historic floods worldwide.

6,940 BC, the Age of Cancer, marked the eruption of Mount Vesuvius in Italy.

6,880 BC, the Age of Cancer, saw the eruption of Mount Erciyes in Turkey.

6,600 BC, the Age of Cancer, marked the Karymsky eruption in Russia.

6,500 BC, the Age of Cancer, witnessed the cultivation of peas as Egypt's second crop.

THE GIZA PARK PROTOCOL HOW TO START THE SUN

6,440 BC, the Age of Cancer, saw the eruption of Kurile in Russia.

6,200 BC, the Age of Cancer, marked significant volcanic activity at Sakurajima in Kyushu, Japan, Karapinar Field in Turkey, and Mount Hasan Dagi in Turkey.

6,060 BC, the Age of Cancer, saw the Haroharo Caldera eruption in New Zealand.

6,050 BC, the Age of Cancer, witnessed the eruption of Menengai in Eastern Kenya.

6,000 BC, the Age of Gemini, marked the eruption of Mount Etna in Italy.

5,900 BC, the Age of Gemini, witnessed the formation of Crater Lake from Mount Mazama in Oregon, USA.

5,700 BC, the Age of Gemini, saw the Khangar eruption in Russia.

5,677 BC, the Age of Gemini, marked another eruption of Crater Lake Mount Mazama in Oregon, USA.

5,550 BC, the Age of Gemini, saw eruptions at Tao-Rusyr Caldera in the Kuril Islands, Russia, and Mashu, Hokkaido, Japan.

5,250 BC, the Age of Gemini, marked the eruption of Mount Aniakchak on the Alaska Peninsula, USA.

5,060 BC, the Age of Gemini, witnessed the eruption of Tuhua on Mayor Island in New Zealand.

5,000 BC, the Age of Gemini, marked the end of the historic floods.

4,750 BC, the Age of Gemini, saw the eruption of Mount Hudson in Cerro, Chile.

4,360 BC, the Age of Gemini, witnessed volcanic activity on Macauley Island, New Zealand.

THE GIZA PARK PROTOCOL HOW TO START THE SUN

4,350 BC, the Age of Gemini, marked the Kikai Caldera eruption in the Ryukyu Islands, Japan.

4,000 BC, the Age of Taurus, saw the eruption of Pago in New Britain. This period also introduced Sanatana Dharma, the first one-world religion, also known as the Vedic Religion or "the eternal religion."

3,580 BC, the Age of Taurus, marked eruptions at Haroharo Caldera in New Zealand and Talisay Caldera on Luzon Island, Philippines.

Chapter Four:
Earth's Cosmic Connection

Contemplating the energies that form the universe's electrical grid, I had a profound realisation. Humans are electrically connected, much like batteries in a series powering an electric golf cart or automobile. By identifying and isolating each of these systems, I uncovered the unique influences of each module—not just on our planet, but on our species, humankind.

Once I identified and isolated these significant systems, I created a working diagram, a blueprint, arranging them into a functional layout model. While still a work in progress, this model provides a solid foundation for my hypothesis: the "Earth's Cosmic Connection."

My aim is to offer a fresh perspective and direction, encouraging fellow researchers and archaeologists to contribute to this collective effort.

Before explaining how humanity's energetic relationship integrates cosmically with the Earth and beyond, I must begin this discussion with a detailed systems analysis of how I believe our solar system functions. I start with the "Earth's Cosmic Connection." As an added benefit, you will discover why I believe the Mayan Calendar ended on 21 December 2012, and why the Earth's magnetic field is shifting eastward, away from the Magnetic North Pole as we have traditionally known it.

Looking at the diagram of the Earth's Cosmic Connection, you will note that the Earth's rotation is anticlockwise—essentially a negative (-) rotation.

However, something not taught in schools is that the Earth's core, composed largely of iron and nickel at the Sun's surface temperature of around 10,000°F, rotates in the opposite direction to the Earth's exterior—a clockwise, positive rotation.

How is this possible?

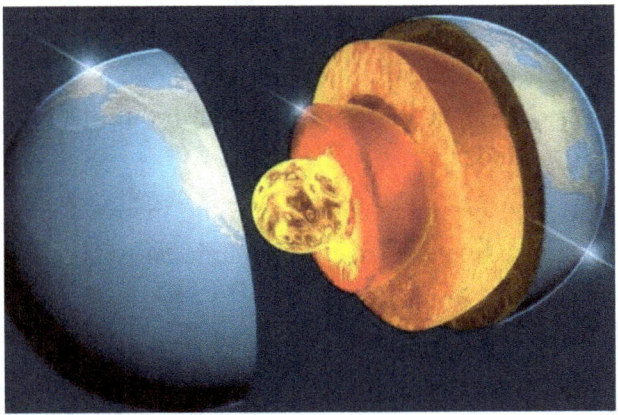

The Sun's heat does not penetrate the Earth's core—if it did, we would all be incinerated.

THE GIZA PARK PROTOCOL HOW TO START THE SUN

Looking at my diagram, I believe the Earth's core receives energy from our "mother sun," Sirius, causing it to rotate clockwise. I also theorise that as Sirius approaches Earth, the rotation of the core accelerates. This phenomenon, I believe, is responsible for the North Magnetic Field shifting eastward at a rate of 50 miles per year. In fact, this shift necessitated the construction of a new runway in Alaska, as the GPS-based landing software could no longer safely guide aircraft to land.

I propose that this increased core rotation speed is heating the Earth's interior, leading to a rise in volcanic activity, more frequent and severe earthquakes, and significant land separations—particularly in California and Africa.

While this is not the primary focus of this volume, it must be addressed so that people can understand what is being withheld from us—because they do not want us to know. The more ignorant we remain, the more control the arrogant bastards have over us all.

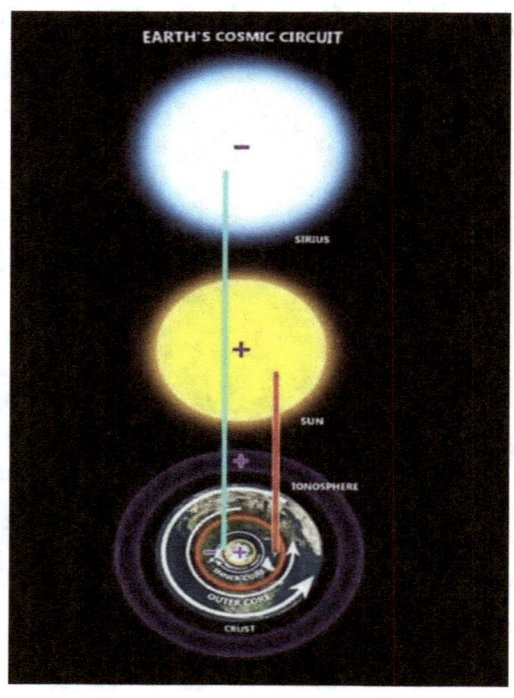

Now, let us delve deeper into this complex discussion of the "Earth's Cosmic Connection." Astronomy is one of the oldest natural sciences, documented across ancient civilisations. I would like to begin with an introduction to an age-old symbolic relic, the "Vesica Piscis," to explore its meaning and how our solar system operates within a cyclical, bipolar relationship between the Sun and Sirius—the brightest star in the night sky, and what some ancient cultures regarded as our "mother sun."

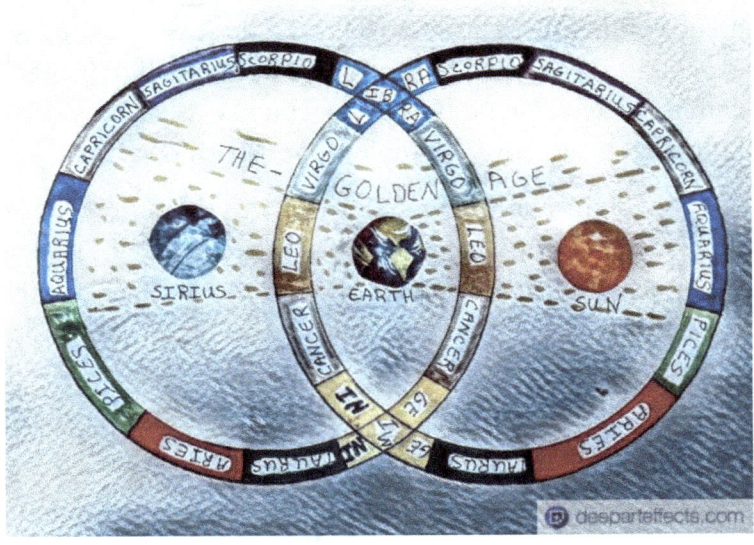

My research suggests that our solar Sun and Sirius, with their profound influence on Earth, form the foundation of an Indigenous Trinity—one that underpins many religious beliefs.

The ancient Eastern Vedic astronomy system has deep historical roots in the northwestern regions of South Asia and has been known by several names, including the Indus Valley Civilisation, the Harappan Civilisation, and the Indus Civilisation. Archaeologists suggest this culture dates as far back as 3300 BC and was one of the great Bronze Age civilisations.

However, as a serious investigator—just as with my research into ancient Egypt—I believe this timeframe could be off by thousands of years, and that the ancient Indian civilisation is far older than commonly accepted.

I am convinced that the Ancient Indians predate the so-called "Old Kingdom" Egyptian civilisations, challenging the widely accepted but misleading notion that Egypt was the "Cradle of Civilisation." There is no definitive proof—only fabricated, elaborate stories and Hollywood-scripted history. Fake, fake, fake.

Additionally, we must acknowledge a physical reality: the Indian subcontinent was once attached to the east coast of Africa before breaking away and drifting, eventually colliding with the Eurasian tectonic plate to form the Himalayas. This raises the possibility that even India itself may have African origins.

As this civilisation matured over the centuries, it began studying and documenting the science of astronomy. One key aspect we should focus on is the natural cyclic relationship they discovered between our Sun and Sirius. They named this natural cosmic cycle "The Great Year."

While the Eastern Indians were developing their methods, the ancient Sumerians created their Cuneiform script, and the ancient Egyptians—beginning in the Upper Nile region—also started their astronomical studies. Numerous other civilisations, including the Mesopotamians, Chinese, and Babylonians, began documenting astronomical observations during the Bronze Age.

However, based on most of my research, I have chosen the Indian Vedic description as the foundation of my discussion. I have created a detailed graph best described as the "Great Year." This graph is progressive and can be considered a hybrid, incorporating some of the so-called "Ages," such as the "Golden Age," as well as the 2,000-year astrological periods, including the "Age of Aries."

THE GIZA PARK PROTOCOL HOW TO START THE SUN

Before we continue, a brief description of Sirius is necessary. According to Wikipedia, the name Sirius originates from the Greek word Sirios, meaning "lit," "glowing," or "scorching."

Sirius is designated as part of the Canis Major constellation, earning the nickname the "Dog Star." It is one of the hunting hounds accompanying Orion, the Hunter. Sirius has an apparent visual magnitude of -1.46, making it almost twice as bright as the second-brightest star, Canopus, and 25 times brighter than our Sun.

In Ancient Egypt, the heliacal rise of Sirius, coinciding with the Sun, marked the annual flooding of the Nile. The Ancient Greeks, meanwhile, associated its appearance with the "Dog Days" of summer.

Many ancient cultures considered Sirius to be our "Mother Sun." They believed it was the centre of our universe, meaning the Milky Way Galaxy system revolved around it.

A significant issue, however, lies in the way ancient civilisations documented this cycle—known as the "Precession of the Equinoxes" or "The Great Year." Many named this cycle in reference to the binary relationship between our Sun and Sirius, yet their calculations contain what I believe to be a serious flaw in the timeline.

Hipparchus, a Greek astronomer, had access to centuries of Babylonian and Greek records. He is credited by the Greeks with discovering the cycle of precession. By comparing ancient records with his own observations, he drew conclusions that have influenced our understanding of this cycle.

Now, here comes the key issue—the flaw I mentioned earlier—that must be addressed before proceeding. My research into "The Great Year" has been extensive, and I have recently made an important discovery.

THE GIZA PARK PROTOCOL HOW TO START THE SUN

For thousands of years, the knowledge we have inherited about Sirius's position has always been based on a specific 13,000-year period—half of the supposed 26,000-year cycle—during which Sirius has been moving away from Earth.

Thanks to the groundbreaking research of Swami Sri Yukteswar, published in 1894 in The Holy Science, we now know that the early Indian sages miscalculated the time frame of the Great Year by approximately 2,000 years. Yukteswar determined that the full cycle is actually 24,000 years, not 26,000.

This means that in the past 12,000 years, we only have documented records covering roughly 6,000 years. As a result, no ancient civilisation has ever recorded a period when Sirius was moving towards Earth. This means that no civilisation has directly experienced the potential energetic influence of our "Mother Sun" when it was closest to our planet. Without this direct experience, any discussions regarding Sirius's impact on Earth during such a phase remain purely theoretical.

One civilisation that may have sensed this shift—without knowing how to prepare for it—was the ancient Mesoamerican and Central American Mayan civilisation.

Let's examine their behaviour. The Mayan civilisation lasted over 3,000 years and used multiple calendar systems to track time and agricultural cycles.

This is widely evidenced by the great Mayan astronomers who concluded their three-calendar system on the winter solstice, 21 December 2012.

For decades, many interpreted this date as a prediction of the world's end due to the absence of a continued Mayan calendar system.

I argue otherwise. Today, we are more than a decade past that date, and the world has not ended. However, my research suggests that 21 December 2012 marked a critical astronomical event: the moment when Sirius reached its farthest precession point from Earth.

Beginning on 22 December 2012, Sirius changed its direction by 180 degrees—shifting from moving away from Earth to moving towards it.

It is highly probable that the Mayan civilisation, which thrived for 3,000 years, had no way of predicting how this course reversal might impact Earth. This uncertainty may have influenced their decision not to create a new calendar system beyond that point.

Even today, in our modern era, we still lack the knowledge to predict the long-term effects of Sirius's shift. Could its approach bring significant changes or even catastrophic consequences for Earth? We simply do not know.

This ongoing mystery only deepens the intrigue surrounding this cosmic relationship.

THE GIZA PARK PROTOCOL HOW TO START THE SUN

The ancient Dogon tribe in Africa predates the earliest Egyptian dynasties. They referred to Sirius as our "Mother Sun." The Egyptians later followed suit, naming one of their most powerful deities, the goddess Sopdet, as the first to personify Sirius.

Sopdet was later associated with Isis, and the two goddesses were sometimes conflated or merged. The Egyptians believed that the reappearance of Sirius, the brightest star in the southern sky, signalled the start of the new year. This event became known as the heliacal rising of Sirius.

It is important to note that Isis became the later dynasty's representative of Sirius and is often depicted with a five-pointed star. Throughout the rest of this text, Isis will symbolise Sirius. I wanted to highlight the historical progression of symbolism in Egypt. When a civilisation lasts for 3,000 years, many powerful rulers come and go, often with absolute control over their domains. As they were considered gods on Earth, they frequently altered the composition of their pantheon.

Within the Egyptian hieroglyphic system, Isis is represented with a five-pointed star on her head. Almost every time a five-pointed star appears in Egyptian hieroglyphs, it can be interpreted as relating to Sirius in some way.

No culture appears to be more connected to Sirius than the ancient Egyptians, particularly in their emphasis on the feminine (-) negative sun across various aspects of their belief system. One of the most critical influences attributed to Sirius—central to my theories on this topic—is its connection to the start of the rainy season each summer.

The heliacal sunrise occurs when the two suns, Sirius and our Sun, rise together in the eastern sky at dawn on the summer solstice. Both align directly with the Giza Plateau.

THE GIZA PARK PROTOCOL HOW TO START THE SUN

The heliacal sunrise took place directly in front of the Great Sphinx, which explains why the Sphinx faces east. For thousands of years, she has watched this miraculous event unfold in a continuous cycle. It is believed that this heliacal sunrise marked the beginning of Egypt's annual rainy season.

Given the profound significance of Sirius's relationship with Earth, now is the perfect time to explore its historical ties with ancient Egypt and the heliacal sunrise.

The heliacal sunrise always occurred within a short window between the summer solstice and the Lion's Gate Portal each year.

Identifying Sirius as the second sun of our binary solar system requires extensive evidence and analysis to fully grasp this concept.

This second sun follows a vast elliptical orbit similar to our own Sun's but moves in the opposite direction. It has a cyclic relationship with our Sun and Earth, affecting the intensity of its energetic influence on our Sun, all planets within our solar system, and life on Earth, including humans and all other species.

To see this second sun with the naked eye, one must look for it at night. Since the Greeks named it, it has been known as Sirius. In astronomy, it is classified as Alpha Canis Major. It is commonly referred to as the Dog Star, as it is the largest of the two dogs following the Orion constellation, the Hunter.

Technical and Historical Data on Sirius

Some of the following information is common knowledge, while some may seem unbelievable.

First and foremost, Sirius is the brightest star in the night sky. It is twice the size of our Sun, 25 times more luminous, and exceptionally hot. Its surface temperature is approximately 9,400°C (17,000°F), compared to our Sun's 5,500°C (10,000°F).

THE GIZA PARK PROTOCOL HOW TO START THE SUN

The Greeks gave Sirius its modern name, and, yes, it represents the feminine negative (-) sun energy in contrast to our positive (+) sun, the Yang. Together, they form the bipolar nature of our binary solar system.

However, around 3,000 BC, everything began to change. Seasonal rainfall declined, leading to the Dry Period, which was declared in 2,500 BC. This shift marked the beginning of the end for the Old Kingdom of Egypt. Without regular seasonal rains, crops could no longer grow in the desert. By 2,200 BC, mass starvation and disease caused widespread civil unrest.

Compounding this turmoil were unstable leadership transitions and devastating external invasions—first by the Sumerians, then by the Persians. These attacks ultimately led to the downfall of ancient Egyptian civilisation, striking at their weakest moment when the rainy season had come to an end.

The Ancient Hindu Perspective on Sirius

Another culture that recognised Sirius's significance long before the Egyptians—and possessed the most extensive scientific records detailing its cycles—was the ancient Hindu Indians.

My journey into this knowledge began with a year-long training programme to become a 200-hour certified Yogi in the Integral System of Yoga.

My guru, Swami Satchidananda Saraswati, modified this method and introduced Integral Yoga, originally founded by Sri Aurobindo, when he established his ashram in Pondicherry, India, in 1926. In 1966, Satchidananda brought Integral Yoga to the United States, founding the first Integral Yoga Centre in New York.

Three years later, Sri Satchidananda became famous in America for opening the Woodstock Music Festival in 1969. To add some Indian influence to the event, the legendary sitar master Ravi

Shankar performed three songs with his classical Indian instrument, the sitar.

This deep connection with yoga opened many doors for me, allowing me to explore ancient philosophy and Vedic science. It was during this period that I learned about the well-documented Holy Science of the Eastern Indians, which originates from the prehistoric and oldest religion on Earth—Sanatan Dharma.

Sanatan Dharma eventually evolved into what is now widely recognised as Hinduism. This term was given by the Persians, meaning "the people who live next to the River Sindhu". Many religious sects and philosophical schools emerged from this foundation, including Buddhism and Jainism.

Additionally, major religions that exist today—including Christianity, Catholicism, Islam, Sikhism, and Judaism—all originated from the principles of Sanatan Dharma. Regarding Judaism, it is important to note that the Jewish people were originally from Egypt before migrating to form Israel. Their religion, and even the name of their nation, is deeply rooted in Egyptian knowledge and heritage. (More on this topic later.)

The Hindu "Great Year" and Sirius

During my year of hands-on yoga training, I discovered another significant cycle recognised by the ancient Hindus: the "Great Year."

According to Hindu tradition, this Great Year is directly linked to the cycles of Sirius and our Sun. It is mentioned in the Mahabharata, one of the oldest and most revered texts in Indian history.

The Mahabharata consists of 18 volumes, including the Bhagavad Gita, one of the most profound spiritual teachings ever written. No other book has had a greater positive impact on my life.

THE GIZA PARK PROTOCOL HOW TO START THE SUN

I have read over a dozen translations of the Bhagavad Gita, including one written by Mahatma Gandhi. However, my personal favourite is The Essence of the Bhagavad Gita, as explained by Gandhi's guru, Paramahansa Yogananda, the man who introduced Kriya Yoga to the West. I have read this version at least ten times, and it has healed and enlightened me in countless ways. It remains the most extraordinary book I have ever read.

Staying focused on this "Great Year," I discovered that Hindu science had investigated and understood this concept better than many modern scientists today. They developed an impressive analogy of the cycles of Sirius interacting with our Sun and how these cycles influenced the Earth and humankind.

For most readers, this may be an unfamiliar subject, so let's delve deeper. The "Great Year," depending on the calculation method used, has some variation in length. However, the most commonly accepted estimate is approximately 26,000 years. Our Sun is positive (+), while Sirius is negative (-) in polarity. Though these two Suns are directly related, their common denominator is their shared influence on the Earth and humanity. According to the ancient Eastern Indians, the "Great Year" lasts 26,000 years. Given this vast timeframe, it is easy to understand why modern humans have little knowledge or connection to Sirius compared to our relationship with the Sun.

More recent research, conducted using advanced measuring instruments and supercomputers by my colleague Walter Cruttenden, founder of the Binary Research Institute, suggests that the "Great Year" is closer to 24,000 years rather than the 26,000-year estimate of the Hindus. This means Sirius's elliptical orbit takes 12,000 years moving away from our Sun and Earth, followed by another 12,000 years returning towards us. As Sirius moves closer, its influence on Earth grows stronger and more intense.

THE GIZA PARK PROTOCOL HOW TO START THE SUN

It is astonishing to look back at the predictions made by ancient Eastern Indians thousands of years ago. They theorised the 26,000-year cycle of the "Great Year" without modern technology or direct empirical evidence to confirm their calculations. Yet, modern measurements have verified their predictions with over 92% accuracy—an extraordinary achievement.

This level of precision in ancient Vedic science is remarkable. It becomes even more intriguing when considering that the Mayan civilisation ended its three-calendar system on 21 December 2012. When I incorporate my current hypotheses into their ancient beliefs, the pieces align, allowing me to present my theories with greater coherence.

I aim to provide sufficient collaborative evidence to support my theories, giving the reader a broader perspective. Why? Because my forensic investigative research, spanning multiple disciplines, has revealed compelling connections between these ancient sciences and our understanding of Sirius today.

Less than a decade ago, when the Mayan calendars ended, many believed the world would come to an end on 21 December 2012. Clearly, that was not the case. However, my years of research suggest that the Mayans may have ended their calendar at that specific date due to a lack of further information.

I propose that both the Mayans and the Hindus based much of their cosmic knowledge on the elliptical orbit of Sirius and its influence on Earth. If we consider that the Mayan civilisation flourished for approximately 3,000 years and went through multiple periods of dissolution and reformation, their knowledge of Sirius was inherently limited to the part of its cycle when it was moving away from Earth.

The Mayan timeline is commonly divided into three main periods:

THE GIZA PARK PROTOCOL HOW TO START THE SUN

- Pre-Classic (2000 BCE – 250 CE)
- Classic (250 – 900 CE)
- Post-Classic (900 – 1519 CE)

The accuracy of these timeframes is based on modern measuring techniques and remains an educated estimate.

The Mayans likely ended their calendar system on 21 December 2012 because this was around the time Sirius reached the farthest point in its 12,000-year cycle before beginning its return towards Earth. Their entire civilisation developed while Sirius was moving away, meaning they had no direct knowledge of its effects when it was approaching. Their understanding was based on a one-sided perspective, lacking historical records of Sirius's increasing influence as it returned.

This is the primary reason I believe the Mayans chose to end their calendars at this specific point in time.

I am convinced that the Mayans, with their extensive knowledge of various sciences—astronomy being one of their greatest strengths—based their calculations on observable influences of Sirius. I must conclude that they, like the ancient Hindus, Egyptians, and possibly even the earlier Dogon tribe, understood parts of the "Great Year" cycle of Sirius.

My forensic investigation experience, spanning over 2,500 cases, has given me deep admiration for the Mayan civilisation. I have repeatedly observed numerous similarities between the ancient Egyptians and the Mayans, and I will highlight these findings throughout my work.

One striking similarity between these two civilisations, which were separated by vast distances, is their use of multiple calendars for different purposes. The Mayans had three distinct calendar systems, just as the ancient Egyptians had separate calendars for various seasonal cycles.

THE GIZA PARK PROTOCOL HOW TO START THE SUN

Another intriguing connection between ancient civilisations is the existence of extensive underground communities across multiple continents. For example, in Cappadocia, central Turkey, there are at least 36 underground cities, some reaching depths of 60 metres (200 feet). The deepest, Derinkuyu, extends nearly 85 metres (280 feet) and could accommodate over 20,000 people, along with their livestock, food, and water stores. These cities included schools, medical facilities, places of worship, and infrastructure necessary for prolonged underground habitation.

Given this, it is reasonable to consider that at certain times in history, the Earth's surface may have been uninhabitable due to extreme weather or climate conditions rather than simply the threat of raiders.

During my two-and-a-half-month stay in Turkey in 2019, I personally experienced the persistent issue of earthquakes. A severe earthquake struck a northern community, causing major destruction and loss of life. About a week later, the hotel where we were staying was violently shaken, and all the residents ran into the streets in panic, urging us to evacuate.

With this in mind, it is essential to consider potential causes of mass extinctions that could have driven ancient civilisations underground. Some possibilities include:

- Massive volcanic eruptions
- Flood basalt events
- Asteroid impacts
- Global cooling or warming
- Anoxic (oxygen-depleted) events
- Hydrogen sulphide emissions
- Oceanic overturn events

- Geomagnetic reversals
- Tectonic plate shifts

By exploring these factors, we may uncover deeper insights into why certain ancient civilisations developed such extensive underground infrastructure and how their knowledge of cosmic cycles, such as the orbit of Sirius, played a role in their survival strategies.

These events collectively impact various ecosystems, but their aftereffects can be devastating. Particles in the air can block sunlight for extended periods, toxic gases in the atmosphere can poison the air, and acid rain can destroy crops. In the oceans, water quality deteriorates, making it uninhabitable for marine life. With rapid temperature changes, both plant and animal life struggle to survive.

Given these factors and our understanding of history, it is crucial to consider the latest scientific studies, which indicate that Earth has experienced five major extinction events. These events wiped out vast numbers of species—not just the dinosaurs. Now, scientists warn that we are facing a sixth mass extinction.

The most catastrophic of these events eradicated 96% of all marine species and 70% of all land species. Birds were among the few to survive. It took millions of years for the Earth to recover from such devastation.

We must also remember the last Ice Age, which lasted 2.4 million years, profoundly shaping the planet and its ecosystems.

Chapter Five:
The Odyssey Begins

I want to welcome everyone aboard my vessel, the SEA SPIRIT, and thank you for joining me, Captain Leo, as your dedicated guide on this extraordinary journey. My Odyssey—a voyage of historical discovery—has been shaped by 15 years of exploration, unearthing mind-blowing revelations.

At present, I am meticulously organising thousands of pages of painstakingly written text, admiring hundreds of original drawings from the many visions I have sketched. I am reviewing dozens of system flow charts, analysing and compiling an extensive body of scientific and theoretical data, including statistical models I have developed over the past decade and a half.

Looking back at the sheer volume of information I have accumulated, I find myself in awe. I pause, reflect, and think, "Wow!" What an overwhelming journey—an odyssey of extraordinary proportions—to finally reach this stage. Now, my greatest challenge is: how do I share these discoveries with the world?

More than just sharing, I aim to reveal a new perspective on our past—one that compels people to clear their minds and, even if just for a moment, open themselves up to the possibility that the history they have been taught is riddled with false narratives and outright lies.

My goal is to present complex ideas and concepts about our cosmos, Earth, and humanity, highlighting how we are all intrinsically and quantumly entangled. I intend to provide such a substantial body of compelling new information that the sheer weight of evidence alone justifies deeper, more aggressive investigation—investigations that must be carried out by those with expertise beyond that of a single researcher and seeker of truth.

So, sit back, get comfortable, and prepare yourself. Forget His Story (history), and allow me to introduce My Story (mystery).

Join me at this remarkable moment in Earth's existence as we hurtle through space at 70,000 mph, racing towards the unknown.

The Awakening

My journey took a dramatic turn on 8 August 2007, a day known as the 8-8 Lion's Gate Portal. Many ancient civilisations, including the Egyptians, recognised and revered this cosmic alignment.

That day, near noon—the very hour the ancient Egyptians would worship Ra, their sun god—I experienced a profound transformation. The Egyptians worshipped five gods each day, from dawn until dusk, but for me, this particular moment would prove electrifying.

I now call it my "Grand Illumination", and since then, I have celebrated 8 August as my second birthday.

At the time, I was in the midst of recovering from a severe traumatic brain injury, the result of an electrocution, a near-death experience, and a sudden Kundalini awakening. It took me a full year to relearn how to read. I could recognise words individually but struggled to comprehend full sentences.

When my ability to read returned, I became insatiable—devouring books, texts, and research at an astonishing pace. I became obsessed with one of the greatest enigmas of all time: the Great Pyramid on the Giza Plateau, Egypt.

A Global Odyssey

THE GIZA PARK PROTOCOL HOW TO START THE SUN

Reflecting on the thousands of miles I have travelled in pursuit of knowledge, I find myself astounded by the scale of my explorations.

Over the years, my wife Heidi Grant and I have visited 12 countries—some multiple times—including Egypt, Turkey, Malta, Spain, India, the United Kingdom, the United States, Mexico, Belize, Greece, and Puerto Rico.

We have examined over 650 sites, meticulously documenting our findings with thousands of photographs and videos. This journey has been an Odyssey of epic proportions, fuelled by an insatiable drive to uncover the truth.

At first, my obsession was singularly focused on the Great Pyramid, but that initial spark soon ignited a volcanic eruption of discoveries, which scattered like breadcrumbs, illuminating a path I followed blindly and willingly—without a map.

I have uncovered so many revelations that I must now produce multiple texts to fully articulate and explain my findings in the necessary detail they demand.

The Giza Industrial Park

As I progressed in my research, I realised that the Great Pyramid was only one component of a much larger system—above and below the Sahara Desert sands and limestone bedrock of the Giza Plateau.

Years of investigation led me to conclude that the Giza Plateau System was part of an even broader network of interconnected Old Kingdom structures.

I coined the term "Giza Industrial Park" to describe this vast system—one powered by brackish seawater aquifers, which played a critical role in sustaining an advanced energy production process. These aquifers, particularly Lake Moeris, provided the essential

THE GIZA PARK PROTOCOL HOW TO START THE SUN

"Mother's Milk"—the fuel that sustained the Great Pyramid's operation.

As my research expanded westward, across Egypt and then Africa, I eventually found myself on the West Coast, in Casablanca, gazing across the vast expanse of the Atlantic Ocean.

I stood there, wondering where my next destination would be—and what new mysteries awaited my discovery.

After uncovering many truths about my homeland of America, I now find myself trying to process those revelations. These truths have been deliberately hidden from us—omitted from our history books, buried beneath layers of manipulation and control, designed to keep us ignorant of the past, oblivious to the present, and without a true future—unless it aligns with their predetermined agenda.

Some of my discoveries may sound far-fetched, like something out of science fiction. Some will be difficult to accept as truth—just as they were for me. They have shattered much of what I was taught to believe, not just about America, but about human history itself across borders and cultures.

Having spent a lifetime in America, I have learned the hard way—through hands-on experience—not to blindly accept anything as fact or truth. America, I regret to say, is nothing like what I was led to believe.

As a child, I was a Cub Scout and Boy Scout, raised to respect my country and everything America stood for. I wanted to defend and protect my nation and my family. Driven by this desire, I became a Nuclear Shipbuilding Apprentice, learning to construct weapons of mass destruction under the belief that I was safeguarding freedom.

I built aircraft carriers, battleships, and submarines, convinced that my work served a noble cause. It took me years to realise how my career had been steered by an agenda I had not questioned.

THE GIZA PARK PROTOCOL HOW TO START THE SUN

For a long time, I have struggled with the karmic impact of my actions—the unintended consequences of my ignorance and the agenda-driven patriotism that shaped my youth. The body count of innocent men, women, and children, lost to false wars and manufactured conflicts, is something I can never erase.

Through painful realisations, I have come to understand that America is not the benevolent nation I was taught to believe in.

A Shift in Purpose

Determined to serve humanity in a different way, I became a professional firefighter, dedicating myself to protecting lives in two separate communities.

However, my deepest obsession has always been the pursuit of truth. This quest led me to Alethiology, the philosophical study of truth, and Epistemology, the study of knowledge and belief systems—why we believe what we do.

With my experience in forensic investigations, my work as a courtroom expert witness, my USCG Captain's Licence, and my role as a founding plank member of Homeland Security following 9/11, I have developed an unyielding commitment to truth.

As a result, I have become far more critical of modern information sources. Since 2008, I have not watched television, listened to the radio, or read newspapers or magazines. I have also abstained from participating in any government voting schemes, recognising them as little more than instruments of systematic deception and theft.

Once you truly understand the depth of deception, you realise just how little truth exists in the world today.

Entangled Theories & Ancient Knowledge

THE GIZA PARK PROTOCOL HOW TO START THE SUN

All my hypotheses and theories are interconnected—or, as I prefer to say, "entangled." There are many ways to acquire knowledge:

- Hands-on experience
- Scientific experimentation
- Intuition
- Extensive research
- Psychic insight

Among the most unexpected gifts of my brain's transformation has been the ability to decode and understand the meanings of ancient symbolic images—across multiple civilisations and diverse historical contexts.

For the longest time, I felt as though my consciousness was drifting through the Akashic Field, tapping into knowledge far beyond my prior understanding.

Even the cover of this text is a symbolic Egyptian hieroglyph, one whose true translation will only become clear once you complete this treatise.

The Anunnaki & The Old Kingdom Systems

Eventually, I identified three primary functions of the so-called Old Kingdom Systems. In doing so, I found recurring patterns that connected back to some of the oldest written records on Earth—those describing the alien visitors from Nibiru, known as the Anunnaki.

According to Zecharia Sitchin's research, as detailed in his work on the Ancient Astronaut Theory, these giant extraterrestrial beings were responsible not only for constructing Göbekli Tepe in Turkey, but also for creating the first pantheon of gods in Sumerian civilisation—with the Anunnaki themselves as deities.

THE GIZA PARK PROTOCOL HOW TO START THE SUN

Thousands of clay tablets, still in existence today, document the Anunnaki's arrival on Earth and their intentions. The translations of these ancient writings contain some of the most fascinating revelations about our distant past.

Among the first significant details I encountered was a description of the conditions on Earth when the Anunnaki first arrived.

Before I go further, let me briefly describe these beings.

The Anunnaki claimed that their purpose for coming to Earth was to mine gold. Their home planet, Nibiru, supposedly required gold to maintain its atmosphere, though the exact science behind this claim remains unclear.

They landed in South Africa and established gold-mining operations, which lasted for decades as they extracted the precious metal to send back to Nibiru.

The most shocking revelation, however, is that the Anunnaki created Earth humans as a slave race, specifically engineered to carry out the physical labour of gold mining.

This is just the beginning of what I have uncovered—truths that have been deliberately erased from our collective memory.

Even the **Old Testament—what I call the Egyptian Bible— **mentions the Anunnaki. The Persians spoke of them, and Alexander the Great credited his global conquest to their legacy. Their influence spans generations long before ours.

The Anunnaki claimed that when they arrived, they found Earth to be a barren, scorched planet, devoid of life and a habitable environment.

I have spent a great deal of time contemplating this description of Earth. My knowledge and forensic experience, driven by an eclectic curiosity and a relentless pursuit of truth, have led me to search for all possible answers.

Looking at scientific data, which has been the largest focus of my research, I discovered that between 200 and 300 million years ago, the Earth consisted of one single landmass—a colossal supercontinent that contained all the present continents joined together. This ancient landmass was called Pangaea.

Now, I have just mentioned that Earth was described long ago as a lonely, single-continent planet, lacking an acceptable habitat for human life.

A Thought Experiment

Imagine I were an outsider, arriving on Earth with the intention of transforming it—creating a thriving, sustainable world, perhaps even a new species capable of surviving and flourishing here. Such an undertaking would be an immense challenge.

To accomplish this, I would first need to:

- Create an atmosphere
- Establish water systems, rivers, and seas
- Introduce an ecological support system
- Separate the landmass into distinct continents and islands

When you consider the sheer magnitude of these requirements, the scale of knowledge and resources needed goes far beyond what humans—as we understand them—could ever achieve alone.

This reinforces my belief that Earth may, in fact, be an experimental planet.

The Moon – Natural or Artificial?

When NASA conducted ground-penetrating tests on the Moon, they used explosives to create shockwaves, which were then measured using specialised equipment. The results led them to conclude that the Moon is hollow.

NASA never explicitly stated this, but based on their findings, I strongly suspect the Moon is an artificial structure—until more conclusive evidence is released.

If the Moon was engineered, then why not the Earth's environment as well?

The Anunnaki & The Creation of Humanity

The Anunnaki also claimed they shared their DNA with a pre-existing species, ultimately creating the first humans on Earth.

According to their records, they landed in South Africa, where they discovered gold. The newly engineered human species were used as slave labour, tasked with mining the gold to send back to Nibiru.

The Evolutionary Timeline – Are We Model 21?

The Smithsonian Institution keeps records of ancient anomalies discovered on Earth. According to their findings, there have been at least 20 earlier human species before us. This means that we—Homo sapiens—are the 21st version of human beings to exist on this planet.

The Agenda for Control

Looking at the modern world, it is clear that governments, corporations, and global organisations such as:

- The United Nations (UN)
- The World Health Organisation (WHO)
- The World Economic Forum (WEF)
- The militarised pharmaceutical and medical industries

...are all working in lockstep to control the global population. And unfortunately, they are succeeding.

The latest tool of control? The mRNA injections—pushed onto the entire population, not for health, but for DNA manipulation.

The Next Human Experiment – Homo Superior Model 22?

Given the agenda to reduce population numbers, I can only speculate that we may be entering the experimental trial period of a new human model—one engineered from those who survive this phase of global transformation.

If this is true, I believe the next human model will be:

- Less intelligent than us
- More biologically modified
- Easier to manipulate and control
- Engineered to be obedient, hard-working slaves

This is exactly what "they" want—a docile, controllable workforce.

A Repeat of History?

Could it be that the same group in power today is using the same methods as before?

Could they be baiting us into volunteering our bodies, minds, and souls for experiments—just as they did with the COVID-19 mRNA vaccines, which, to this day, remain experimental?

The scientific community has already decided the future of humanity. I call it Model 22.

At present, we are known as Homo sapiens – Model 21.

However, our replacements—the next version of humanity—will be called Homo Superior – Model 22.

THE GIZA PARK PROTOCOL HOW TO START THE SUN

An odd coincidence, as my wife Heidi Grant pointed out, is that as Homo sapiens, we have 22 chromosomes. Therefore, I find it both unsettling and exciting that the next version of us—modified to the new 22nd model—aligns with these 22 chromosomes.

The Oxford Dictionary defines Homo Superior as the species that will evolve or be developed from Homo sapiens, with more extraordinary intellect or physical abilities and often possessing paranormal powers.

Let's discuss this definition from a couple of perspectives. Using the same source, the Oxford Dictionary defines "evolve" as developing gradually. Alright, let's develop Oxford's definition further. According to Oxford, to "develop" means to grow or cause to grow and become more mature, advanced, or elaborate.

Now, growing doesn't require much explanation. But there's one more term I want to explore: elaborate. Oxford defines "elaborate" as involving many carefully arranged parts or details, intricate and complicated in design and planning.

I don't know about you, but this last definition should raise an eyebrow. For me, it caused an adrenaline rush, bringing goosebumps. This physical response is triggered by the sympathetic nervous system—the fight or flight reflex—and signifies fear, as it causes the hair on the back of your neck to stand up. And that, frankly, concerns me.

When I think about how fear made it so easy to convince 70.6% of the 8.1 billion people on this planet—that's 5.7 billion individuals—to inject at least one dose of an untested experimental combination of mRNA drugs, which instruct the body's DNA to focus the immune system on COVID-19, weakening it against other invaders, I'm struck by the scale of it all. Currently, 13.53 billion doses have been administered globally, with 9,614 new doses given every day.

THE GIZA PARK PROTOCOL HOW TO START THE SUN

I'm proud to say that I remain an organic human, and I refuse to allow anyone—whether for profit or population control—to inject untested substances into my body. I say this as a member of the Baby Boomer generation, now feeling as if I've been placed on an endangered species list, without any laws to protect me.

I can already see the sales pitch aimed at the unaware: "Oh, come on, inject this chemical into your body to protect yourself for life against the silent, invisible killer—cancer. You know that 50% of the population gets cancer; this tiny injection will protect you, making you a superior human—the first of its kind. You'll be more innovative, faster, more robust, and live a longer, healthier life."

My sister, at 56, became a vaccine advocate—aggressive in her verbal attacks on anyone who wasn't vaccinated, bigoted and threatening. Four months after her second injection, she developed a blood clot in her leg, leading to immediate medical treatment as she struggled to stay alive. Sadly, a few months later, she passed away, leaving behind a husband, two children, and two grandchildren to mourn her, all now burdened with economic debt and the loss of her earned income.

This, along with the mechanisms and laws being implemented—cloud seeding, chemtrails, more chemicals in our food, water, and air, CERN, HAARP systems, and as many toxic substances as they can inject into our bodies—suggests to me that history is repeating itself.

I asked myself whether the Anunnaki are still in control. My answer now is that I don't know for sure, but based on my discoveries about ancient technologies, I don't believe so.

My beliefs about Earth stem from the information I have gathered. I am not religious, and I feel very strongly about the three major religions, which some refer to as warring religions. I call them blood cults, as one was created by the Jews and Romans to establish a War Economic System based on bigotry and hate. Sadly, this inhumane experience has cursed humanity for over 1,000 years. On a positive note, my desire for global peace seems within reach.

However, achieving global peace will come at a cost—human lives, if the current wars continue. The biggest issue in the short term is that these three cults must either transform into peaceful, unbigoted, kind, and accepting communities, or they will be eradicated.

We must ask ourselves: how easy will that be when their entire existence has been built on earning money from hate and killing one another? This, I believe, is why the World Economic Forum and the United Nations want to establish a single approved religion and a global medical programme similar to the National Health Service in the UK.

I was born in a country created for war and the destruction of nations, without regard for human life or concern for our planet and environment.

The small town where I was raised in the early 1950s and 60s was voted the most polluted city in America. The sky was rarely visible. Smoke filled the air, and our lungs were constantly exposed to a stench and harmful chemicals, which also coated our cars and property, requiring frequent washes. The river I grew up beside was black, with yellow foam floating on the surface. I never entered or touched it. The fish in the river were never consumed by us.

Growing up, watching my family and many friends succumb to cancer, I began to realise that Americans have little regard for American lives. If they can ignore the suffering of their own, how can they possibly care about wiping out entire civilisations?

Today, in Virginia, abortion laws allow a full-term pregnant woman, in labour, to have her baby killed legally. The Governor of Virginia, a former gynaecologist, takes pride in making the newborns comfortable before ending their lives.

THE GIZA PARK PROTOCOL HOW TO START THE SUN

I am open-minded and not easily manipulated by others' agendas or sick perversions of control. I firmly support free will and the freedom to examine all information from as many perspectives as possible.

Now, the opportunity exists to share additional information about the Anunnaki's agenda and their claims of creating the first humans through interbreeding.

Scientists have determined that Earth has experienced five extinction events, in which either all or most civilisations were wiped out, leaving a few survivors to carry on and rebuild.

I wonder how often the human model has been erased and replaced with a new one. I believe we are currently undergoing one of these extinction events, especially following the three years of pandemic-driven weakening of humanity. The entire world has been in synchrony, administering DNA-modifying chemicals into millions of bodies. I believe this is part of a darker agenda.

I predict a significant die-off is already underway and will intensify as the Baby Boomer Generation, the Millennials, and Generation X are gradually removed.

The new human will be AI, more docile, and easier to control.

Prepare for an extraordinary adventure into realms that most are unaware of as we explore cosmic phenomena, Earth, human history, pyramids, volcanoes, the Ice Age, and catastrophic floods, all while learning about quantum energies from a new perspective.

As I reverse-engineered every component of the structures above and below the ground on the Giza Plateau, I discovered perfection everywhere I looked. I studied geography, construction, and alignments. Every structure on the Plateau was perfectly aligned, except for three.

The first, the most obvious, is the so-called modern temple, built with its alignment pointing directly at the Giza Sphinx.

The second, the Wall of the Crow, is located closest to the Giza Sphinx and is easily identifiable. It is a megalithic wall with a large hole in it. But is it really only 650 feet long?

The third, which took the longest to identify and was the most difficult to locate, is 108 feet below the ground.

THE GIZA PARK PROTOCOL HOW TO START THE SUN

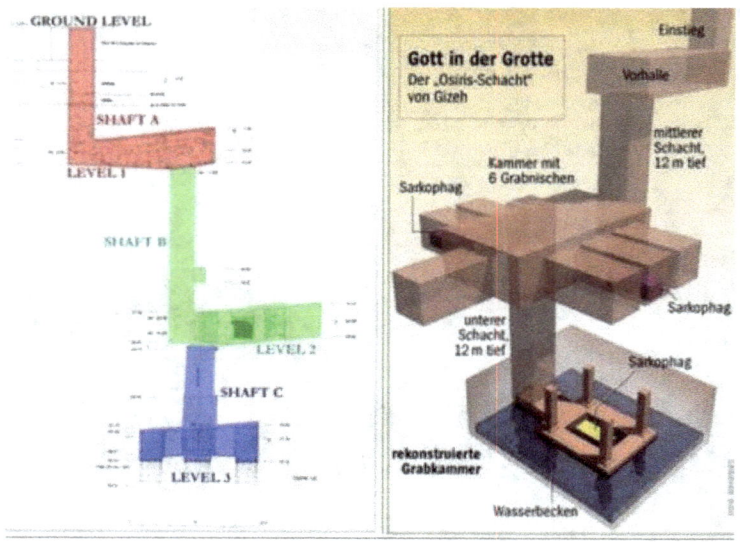

This is Osiris's Tomb, which should be aligned at 90 degrees East, but instead is misaligned at 121.5 degrees.

After extensive research and investigation, I've determined why and how these misaligned anomalies function in a world that otherwise appears perfectly designed. I promise to examine the first two later and will now focus on Osiris's Tomb.

I want to focus on Osiris's Tomb because of its significance—not only to the Supreme Egyptian God, Osiris, Ruler of the Underworld.

Once the water circulates through the Plasma Reflux Pyramid and exits via the Causeway, the newly charged, structured, and heated plasma water (remember, the ionosphere reaches 440°F during the day) drops 35 feet, impacts a limestone stone, and changes direction, sending the waterfall into Osiris's Tomb.

The water then cascades down again, falling 43.5 feet, hitting another limestone surface before pouring into a room with six chambers carved into the bedrock, some containing extra-large, coffin-like stone boxes.

THE GIZA PARK PROTOCOL HOW TO START THE SUN

After the water flows naturally through this maze of sarcophagi, it's clear that this design represents the "Most Advanced Fluid Dynamics" system currently on the planet—and it's ancient.

However, I've uncovered two more such systems to share: one from India and the other from Mexico. But it was Osiris's Tomb that led me to these others.

The plasma ionic waterfall is once again driven by gravity. Torrents rise another 24.6 feet, crashing onto a rudder-designed limestone block, which guides the chaotic fluid to a target inside Osiris's Tomb, now 108 feet deep.

The rudder stone causes the dynamic, flowing waterfall to circulate in a Cyclotron within the natural construct. Various stones with differing polarities, working in conjunction with the powerful magnetic influences from the four cardinal directions, generate an alternating, self-excited longitudinal wave of scalar energies.

Why is it so profoundly carved through the limestone bedrock, with three levels, and a tunnel below? A tomb designed to funnel water from the Ionic Pyramids Causeway, which leads to the Geyser Sphinx.

Itinerary:

Our journey begins at depths of 108 to 116 feet, buried beneath the sands and carved from limestone bedrock, precisely at Osiris's Tomb on the Giza Plateau in Egypt. I invite you to catalogue all the discoveries this volume will bring to light. Enjoy.

Starting now, from Egypt, we will head west, using the bioenergetic anomaly I discovered as our guide—a sort of treasure map directing us to follow a compass heading as we navigate towards unknown places. This will lead us to uncover many treasures of knowledge and wisdom.

This path shows us that the Hall of Records has revealed its secrets and points to 243 degrees west. Our adventure begins with

THE GIZA PARK PROTOCOL HOW TO START THE SUN

me uncovering this mysterious breakthrough—an energy-producing anomaly buried deep below the Giza Plateau.

A clue is presented: An "X" marks the spot, with coordinates reading 121.5 degrees east and 243 degrees west. Using my navigation charts, I eagerly look in this direction, but all I see is desert and sand.

Our journey will be slow and deliberate. As we cross the hot Sahara Desert, you'll come to realise that what was once a lush and green landscape is now a vast, arid expanse—the hottest incidental radiant heat source on the planet.

Steady as we go, we set off, travelling through this scorching desert along the northern coast. Our first point of contact is Tripoli in Libya. From there, we continue our journey 2,285 miles to the west, arriving in Casablanca, located on the coast of the Atlantic Ocean.

Our path: We stand on the shore in Casablanca, looking out over the vast expanse of the Atlantic. We now head across, uncertain of where our treasure map will take us.

With little of note to report from the African continent, I set my sights on crossing the Atlantic Ocean, unsure where this directional anomaly will lead. Like a blind man with a stick, I venture west across the great blue expanse.

Chapter Six: Sea to Shining Sea Tour & More

It wasn't long after I reached the east coast of America, in the state of Maryland, that I entered the Ocean City Harbour. I had never considered the significance of arriving at a place named after Mary until I investigated how the state got its name. We had crossed the vast continent of Egypt with nothing, and the moment we touched solid ground in America, we found something historically intriguing to explore.

As I learned, Maryland was named in honour of Queen Henrietta Maria (1609–1669), wife of Charles I (1600–1649), King of Great Britain and Ireland, who signed the 1632 charter establishing the colony. The region was primarily settled by Roman Catholic immigrants.

My research revealed that Queen Henrietta Maria, being a devout Roman Catholic, faced many bitter struggles in the United Kingdom. The Church of England never crowned her as queen. She was well-versed in the symbolism used by her family for generations. Based on this, I suspect that the name Maria—or Mary, in English—is strongly linked to the Virgin Mary, the mother of Jesus.

Shifting focus to our next discovery, I could never, in a million years, have predicted where our next destination would lead.

We didn't have to travel far before entering the capital of the United States of America: Washington, D.C., a city named in honour of the nation's first president, George Washington.

THE GIZA PARK PROTOCOL HOW TO START THE SUN

I found it fascinating that my journey, in pursuit of an ancient Egyptian god's tomb, was pointing directly at the capital of the United States.

As an American, all I could think was—WTF?

Hold on—this is where things get insane.

Travelling through the heart of the capital, I suddenly stopped in shock, checking my coordinates to verify my course. After confirming them not once but twice, I gasped.

I had landed, with absolute precision, directly atop the Washington Monument—the tallest Egyptian-style obelisk in the world.

THE GIZA PARK PROTOCOL HOW TO START THE SUN

The Washington Monument is precisely 111 miles from the harbour in Ocean City, Maryland.

I must declare that this discovery should—and will—create a great deal of global intrigue, both in America and abroad.

Having found no significant discoveries across the continent of Egypt or the vast expanse of the Atlantic Ocean, our treasure map has now led us to our first remarkable finding:

It is pure, solid gold.

An astonishing 3,800 miles separate the shores of Africa from the pinnacle of the Washington Monument in the capital of the United States.

I find this first major discovery fascinating, as it marks the first connection between the Egyptian Giza Plateau and the ancient replica of an Egyptian obelisk standing in America's capital.

This discovery was a monumental sign for me—confirmation that my treasure map, provided by the Hall of Records beneath the Sahara Desert's sands, was authentic. At that moment, I knew the thousand questions spinning in my mind would soon be answered.

This magnificent quantum electromagnetic scalar cyclotron energy system has opened my eyes to truths long obscured by the lies and deceit woven into the narratives we have been taught.

For the first time since these ancient pyramid systems were constructed and made operational, you are about to witness and understand truths that have been deliberately hidden.

As I write this, Heidi and I have just completed a four-month investigative adventure, which we named our "Sea to Shining Sea Tour." We chose this name based on a lyric from the well-known American song *America the Beautiful*, an idiom signifying a journey from the Atlantic Ocean to the Pacific Ocean.

THE GIZA PARK PROTOCOL HOW TO START THE SUN

You will be amazed by the discoveries we made during this extraordinary odyssey—an expedition aimed at reshaping history and advancing scientific understanding. Our research, built upon a foundation of collaborative inquiry and investigative spirit, has revealed an abundance of new information.

Incidentally, *America the Beautiful* shares its melody with the British anthem *God Save the Queen (King)*. While this journey reflects the geography referenced in the song, the treasures we have uncovered extend far beyond its lyrics.

Since we are here, let's explore some critical details we have learned about Washington, DC.

First, I want to describe the layout of the National Mall. This is where the government has constructed memorials dedicated to those they want you to remember—primarily war memorials, tributes to assassinated presidents, and one honouring a civil rights leader.

Remember, America prides itself on its military might, maintaining 250 bases worldwide as a display of its power.

The Washington Monument stands at the centre of the National Mall, perfectly aligned with the Osiris Tomb in Egypt and several other significant monuments.

One particularly peculiar feature of the Washington Monument—named after the first US president—is that it is the world's largest Egyptian-style obelisk. This towering structure sits within intersecting circles that form the well-designed Vesica Piscis symbol.

(See images.)

The Vesica Piscis symbol has been associated with Christian traditions for a very long time. However, if you were to ask most Christians, they likely wouldn't have a clue what you were referring to.

In Christian art, Jesus Christ is often depicted inside the Vesica Piscis oval, which symbolises the Virgin Mary—the Mother Goddess—and her womb.

The Vesica Piscis also represents the Divine Feminine. It is sometimes compared to the vagina or vulva, allegorically referred to as the "womb of the universe"—a symbol of fertility in both Christian and Pagan belief systems. This imagery is also prevalent in Catholic traditions.

The Vesica Piscis symbol at Chalice Well in Glastonbury, UK, is closely linked to Pagan traditions.

Let's take a moment to review some intriguing details.

THE GIZA PARK PROTOCOL HOW TO START THE SUN

What I call the "Osiris Line" has so far led us to a state in America named after Mary—a symbol of fertility. The world's largest replica obelisk stands in the capital of the United States, and this massive structure is placed within the Vesica Piscis, a feminine symbol representing fertility.

Now would be a good time to briefly explain the mythological story of Osiris so that everyone can understand the interpretations I will offer later. Soon, I will also delve into a more technical discussion on plasma, so I must first approach the topic from different perspectives—connecting ancient allegorical symbols with their deeper meanings.

To make this easier to understand, let's start with the Egyptian myth of Osiris. He was murdered by his jealous and vengeful brother, Seth, who sought to take over his rule.

Ancient Egypt was a civilization where royal bloodlines often resorted to murder to maintain or seize power. Infighting and brutal assassinations were common among ruling families. Seth and Osiris had a long history of rivalry, and tensions escalated when Osiris made Seth sterile—eliminating his chance to produce an heir and establish a lineage.

Determined to exact revenge, Seth devised a plan to destroy Osiris's legacy before he and his wife, Isis, could conceive a child. With calculated cruelty, Seth carried out his brother's murder in a manner that was not uncommon in Egypt. Even Cleopatra, centuries later, convinced Mark Antony to execute her 16-year-old sister, Arsinoe.

Seth then dismembered Osiris's body, cutting it into 14 pieces. As he traveled across Egypt, he scattered these fragments— including Osiris's genitalia—so that his brother could never be reassembled, even through the great alchemy and magic of Thoth.

THE GIZA PARK PROTOCOL HOW TO START THE SUN

In desperation, Osiris's wife, Isis, and her sister, Nephthys, searched the land—through swamps, mountains, and deserts—until they recovered 13 of the 14 body parts. Missing only Osiris's genitalia, Isis pleaded with Thoth to create a phallus so she could conceive a child.

Thoth successfully crafted a magical phallus, and Osiris—God of fertility and the archetype of the dead and resurrected king—was able to father a son with Isis. Using her divine abilities, Isis transformed herself into a beautiful kite bird and became miraculously impregnated. She later gave birth to their falcon-headed son, Horus, the God of the Sky.

The Vesica Piscis also has astronomical significance. It represents a celestial cycle—the relationship between our Sun and Sirius, known to the ancients as our "Mother Sun." This symbol reflects the orbits of these two stars and is linked to what Hindu Vedic sciences call the *Great Year*, a repeating cycle of 24,000 years.

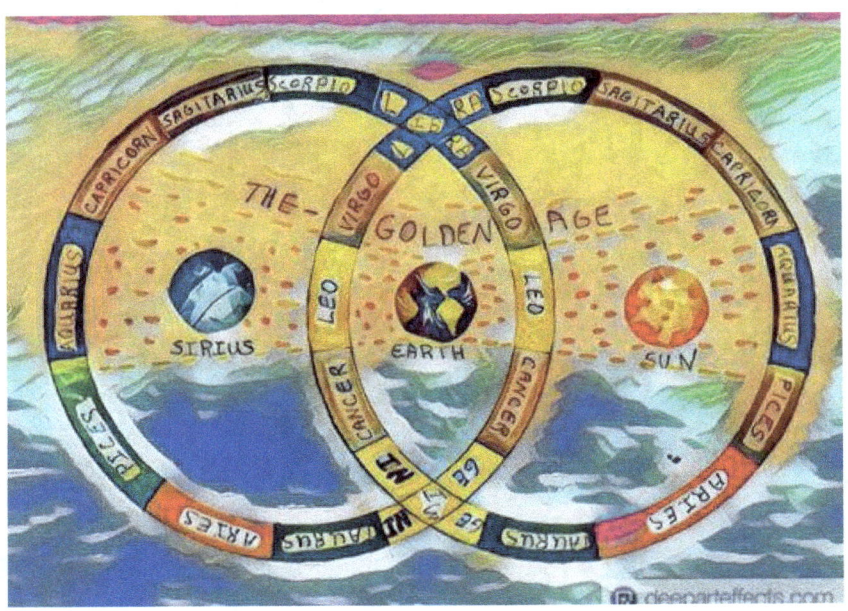

THE GIZA PARK PROTOCOL HOW TO START THE SUN

Now that we have built upon new information, we can return to the National Mall in the capital of the United States.

Here, we find the Abraham Lincoln Memorial, positioned in a remarkably precise alignment with the Washington Monument. Lincoln also shares an intriguing relationship with the Thomas Jefferson Memorial, the Martin Luther King Jr. Memorial, and the gravesite of another assassinated U.S. president, John F. Kennedy.

Take note: the pyramid alignment extends from the Thomas Jefferson Memorial to the Lincoln Memorial, crossing over the Martin Luther King Jr. Memorial. Soon, you will understand the significance of this alignment.

See the image. This is the first time I will introduce the concept of pyramid—or volcano—alignments among these significant sites.

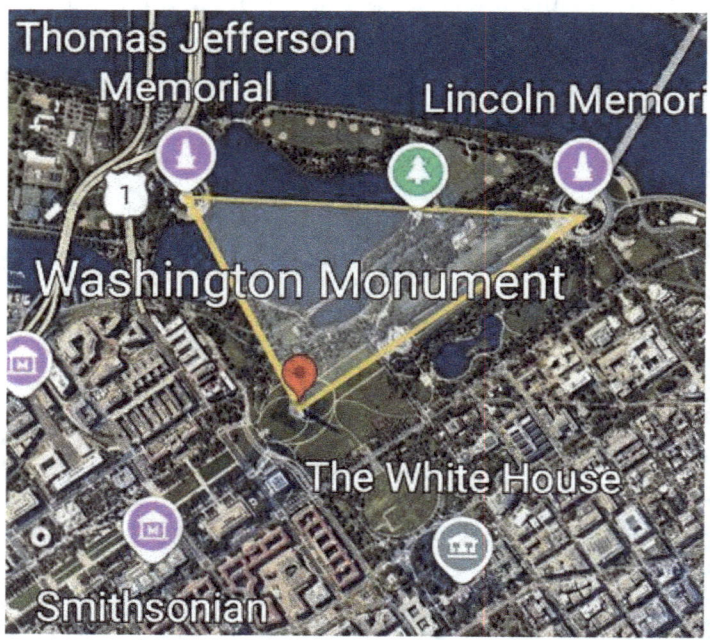

THE GIZA PARK PROTOCOL HOW TO START THE SUN

While this information may not seem immediately significant, it serves as a foundation upon which we can build a "mountain" as our odyssey continues.

The Abraham Lincoln Memorial sits elevated, requiring a climb up steps and a steep incline to enter. Before you reach the steps, two flaming Fasces pillars greet you outside. Once you ascend and witness the powerful image of Abraham Lincoln seated on his throne, flanked by two additional Fasces pillars, it gives me a strange sense of unease. Images of Rome, Mussolini, and Hitler come to mind.

THE GIZA PARK PROTOCOL HOW TO START THE SUN

Abraham Lincoln faces due east, directly staring at the Washington Monument, with the U.S. Capitol Building far behind the obelisk. An intriguing celestial event occurs here daily, which is why the Lincoln Memorial was constructed at this precise location.

Each day, the sunrise rises behind the Washington Monument, allowing Lincoln to witness it. He watches the sun rise from the

THE GIZA PARK PROTOCOL HOW TO START THE SUN

Atlantic Ocean, behind an Egyptian obelisk dedicated to the first President of the United States.

Remember the earlier reference to "America the Beautiful," with the lyric "From Sea to Shining Sea." I believe you can begin to see where this is heading.

Before we move on, I'd like everyone to take a few moments to familiarise themselves with the layout of the National Mall for later comparisons.

(See images.)

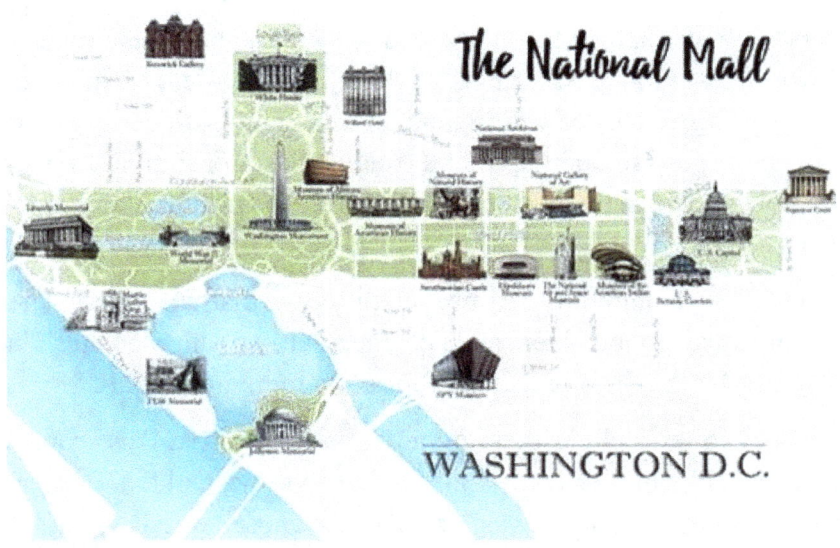

You can easily see the relationship between the Lincoln, Washington, and Jefferson Memorials. Additionally, the location of the Reverend Martin Luther King Jr. Memorial, positioned between Lincoln and Jefferson, is significant. Dr King was a Black Civil Rights leader in America, who was assassinated, much like Presidents Lincoln and Kennedy.

If you exit the Lincoln Memorial from its rear, a dock stretches over the water, leading to Arlington National Cemetery. If you walk 1.25 miles, you will reach the gravesite of Assassinated President

John F. Kennedy. Washington is filled with hidden information that I was never taught.

Leaving Washington, DC, we find ourselves just north of Chicago. Before we discuss the Chicago area, I want to make some interesting observations about the relationship between Egypt and America. Additionally, America has a concerning relationship with Israel. I need to share some unusual comments regarding some of my forensic investigations.

Let's begin with Egypt. It is said that 600,000 Egyptians experienced the Exodus, crossing the Red Sea. These people supposedly worshipped a Pantheon of Egyptian gods at the time, all of whom revered the Sacred Bull. Along their journey, Moses had a change of heart. He decided to establish a monotheistic religion with just one God—something that I find strikingly similar to what Pharaoh Akhenaten did in Luxor a few decades prior to the Exodus.

Akhenaten rejected the Pantheon of gods, much like Moses, and moved his community from Luxor to Amarna, where they worshipped one God, Aten, and became a sun-worshipping cult. The story of Moses mirrors this transformation in many ways.

So, Moses declares that no longer will there be a Pantheon, no more worship of the Bull, and he destroys the Golden Calf. He proclaims himself and his vast group as the chosen ones, demanding the worship of just one God. This shift, however, didn't bode well for Akhenaten and his family, who attempted to do something similar.

Without going into too much detail, Akhenaten's community was attacked with a bioweapon. Yes, even in ancient times, the Egyptians were capable of such cruelty. Akhenaten lost many of his children and his mother to this bioweapon. He was also accused of heresy, and his name was erased from the Egyptian list of Pharaohs.

THE GIZA PARK PROTOCOL HOW TO START THE SUN

How did Moses manage to escape such fate? Perhaps that's why he was constantly on the run. My guess is that there were hundreds of thousands of monotheistic believers. After the destruction of Amarna and the murder of King Tut, the Ramesses initiated an ethnic cleansing, much like the genocides we see today, carried out by the same mindset. Nothing has changed from the barbaric mentality of certain humans who need to be stopped.

What perplexes me about the entire Exodus story is that the new Jewish community seemed to be forming its state by naming their country after three different Pantheons of gods—two from the Egyptian Pantheon and the supreme god of the Canaanites.

The name "Israel" appears to reference three deities from these pantheons. Let's break it down in more detail.

IS-RA-EL: "IS" refers to the Egyptian goddess Isis, who was married to Osiris. Osiris, after being murdered, was resurrected and became the god of the underworld, often referred to as "Ra," the supreme Egyptian sun god. The "EL" part refers to the supreme god of the Canaanites, symbolically associated with the bull or sacred cow. So, despite rejecting the Sacred Cow, the new state worships a deity symbolically linked to it.

IS-RA-EL is, therefore, an Egyptian name for a new nation formed by monotheistic Egyptians and Canaanites.

I've researched Pharaoh Akhenaten, the father of King Tut, extensively over the past 17 years, studying the entire dynasty and the dramatic events of that time. I have a significant issue with accurately identifying the people who left Egypt. Why? Because all those who settled in Amarna mysteriously disappeared. Where did they go? Even Akhenaten and his wife Nefertiti's bodies have never been found.

Considering the documented events from this period, including biological genocide, censorship, the erasure of Pharaohs from history, and the disappearance of their corpses, one must question the details. Were Akhenaten and Nefertiti the parents of the disabled and deformed King Tut, who died at just 19? Egyptians were brutal people—Tut, who had been brought back to Luxor as the capital of Egypt, returned to worshipping the Pantheon of Egyptian gods, abandoning his father's monotheistic religion.

At just 19, King Tut had some strong beliefs. But I feel that he was manipulated by those who destroyed his family and erased his parents from history.

We must also consider the military slaughter of countless tribal communities, particularly those that were monotheistic or had elongated skulls, along with the impacts of drought and famine. The similarities are striking, and it is clear that something is wrong with the details provided.

So, are the Israeli people of Egyptian origin? If you believe the information that is often presented as gospel, I would say yes.

Now, let's address another issue: the barbaric act of circumcision, performed on most newborn males in America. This is an ancient Egyptian blood sacrifice, a ritual performed by followers of the Osiris cult to honour him. We are already familiar with the horrific story of Osiris.

The Egyptians who later settled in Israel still practice this barbaric blood sacrifice today. Some of my Jewish friends have told me that a man cannot be considered a Jew if he is not circumcised. This remains puzzling to me—perhaps this is part of the plan.

We know that the Jews and Romans created the Christian religion. They imposed this same practice of sexual mutilation on Christians, requiring them to undergo this blood sacrifice as penance, in order to be able to worship their God.

Now, let's turn to the relationship between America and the Jewish nation of Israel.

THE GIZA PARK PROTOCOL HOW TO START THE SUN

Akhenaten and Nefertiti were physically different from the other Egyptians of their time. Both had elongated skulls, and Akhenaten, in particular, appeared quite alien. I am sharing this information because my research has been ongoing for 17 years, and I have been aggressive in pursuing these investigations.

During that time, I visited many places around the globe, and one recurring theme was the discovery of people with elongated skulls. One of these investigations occurred in February 2020, when Heidi and I visited Malta. We became intrigued by stories suggesting that the people there had elongated skulls and decided to investigate claims about the Hypogeum site, which was said to be a sacred healing chamber.

What I learned first is that these people, much like the Egyptians and Americans, lie about their history. The entire island of Malta was discovered to have been evacuated, and thousands of people simply vanished. The history lessons provided to the public are disappearing, much like those of ancient Egypt. However, they haven't been as successful in Malta at erasing all the evidence.

Following our investigations at numerous sites on several islands, we eventually made our way to the Hypogeum. The Hypogeum is a sizeable underground excavation the Malians sell as a special healing chamber that vibrates at a healing frequency of 111 Hz.

THE GIZA PARK PROTOCOL HOW TO START THE SUN

I can't help but laugh at this ridiculous claim. For one, they've installed so many walkways, staircases, and fixtures at the site that to suggest there are 111 "healing frequencies" is one of the biggest jokes going. If that's not bad enough, you're not allowed to bring any instruments or cameras into the Hypogeum. This is clearly to keep the pictures out of circulation so that the tourist dollars can keep flowing in.

Our efforts showed us that the online images don't reveal the staircases or walkways.

Moreover, the Hypogeum ended up with an entire housing development built on top of it, so for the longest time, no one knew where it was. When it was eventually rediscovered, they were shocked by what they found, but strangely, evidence of their discoveries started to disappear.

Once they began a closer examination over time, they uncovered 7,000 bodies stacked inside the Hypogeum. These bodies all appeared to have been slaughtered—seemingly the entire population of the island had been killed and left to dehydrate in the sun.

It eventually became clear to me that someone had decided to clear the land of the original Maltese genocide victims. The remains of the murdered people were hidden below ground in this so-called "sacred healing chamber."

THE GIZA PARK PROTOCOL HOW TO START THE SUN

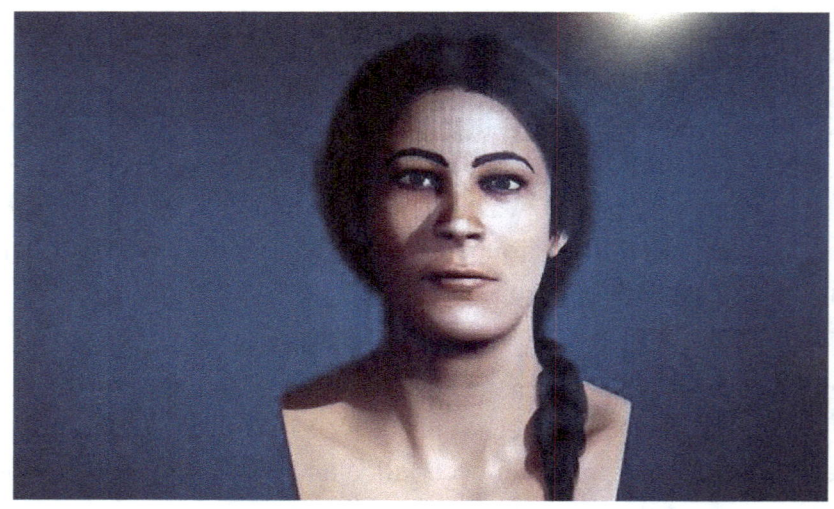

THE GIZA PARK PROTOCOL HOW TO START THE SUN

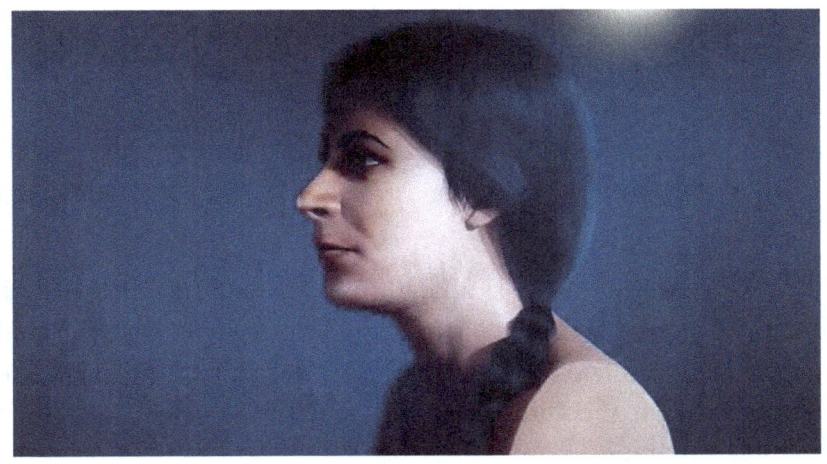

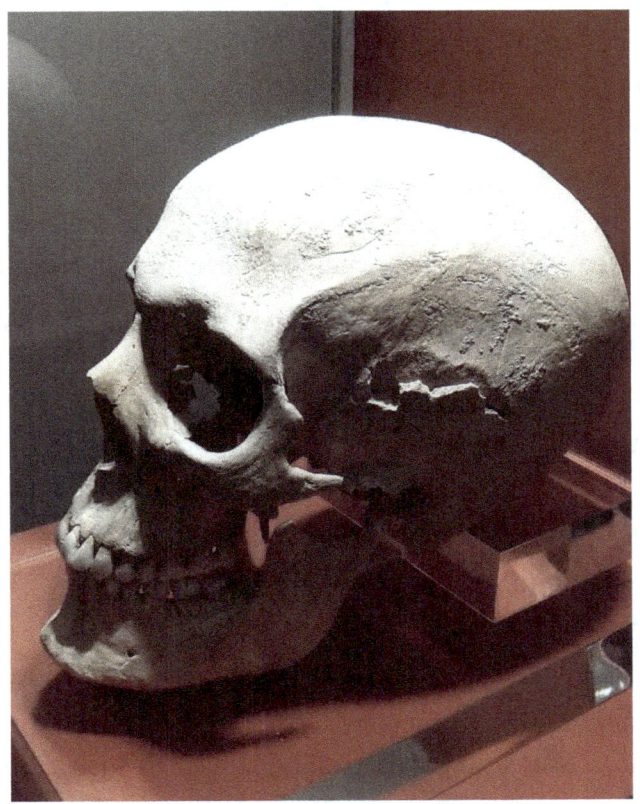

THE GIZA PARK PROTOCOL HOW TO START THE SUN

In our inquiries into local knowledge, we learned that all but two of the skulls of the murdered Maltese had been destroyed and disappeared—twice. Eventually, we located one of the skulls in a museum, which I examined closely. The museum had reconstructed it as the skull of a Mediterranean woman, brown-skinned and attractive.

When I examined the skull carefully (though unable to handle it), I noticed that it had been chemically cleaned, with even the teeth sparkling white. Secondly, the skull was a typical human skull—there was no sign of natural elongation, nor any evidence of artificial cranial deformation.

The only unusual feature was a horizontal break in the skull at the temporal lobe, measuring between 4 and 5 inches in length, just behind the left eye on the left side.

Once we returned from our tour of Malta to London, I immediately began searching for a possible murder weapon that could have caused this injury to the Maltese woman's skull.

During their time, ancient Egyptians studied weaponry, as well as tools for chopping and carving wood. I discovered that the ancients had a popular tool used for cutting, shaping, and moulding wood, particularly in boatbuilding and carpentry. This tool was called an *Adze*. It was a hatchet-style tool with a wooden handle and a bronze metal blade secured to the handle by leather thongs.

Our visit to the Petrie Museum proved fruitful, as we quickly located one of these tools in the museum's extensive collection.

As I mentioned earlier, my primary focus during this visit to the Petrie Museum was to identify a potential murder weapon. The blade of the *Adze* is razor-sharp, and if I had the murdered Maltese woman's skull in my possession, I believe the blade would fit perfectly into the wound site on her skull.

THE GIZA PARK PROTOCOL HOW TO START THE SUN

Following this, I continued my research by examining artifacts at University College London, which had funded and contracted Flinders Petrie's archaeological expeditions to Egypt. The college is also home to the Petrie Museum of Egyptian Archaeology, which houses over 80,000 artifacts spanning from the Pre-dynastic to the Islamic period. These artifacts were sent to the museum following Petrie's discoveries.

The next three images I took at the Petrie Museum in London show artifacts from Petrie's research in Egypt, as I began my forensic investigation. These artifacts could, in my view, have been used as murder weapons in the killings of this young woman—and possibly 6,999 other men, women, and children.

Why do humans engage in genocide? Is this a designed flaw, part of some behavioural experiment? To me, it sounds like the perfect laboratory protocol for ending an experiment and preparing the infrastructure for the next one.

THE GIZA PARK PROTOCOL HOW TO START THE SUN

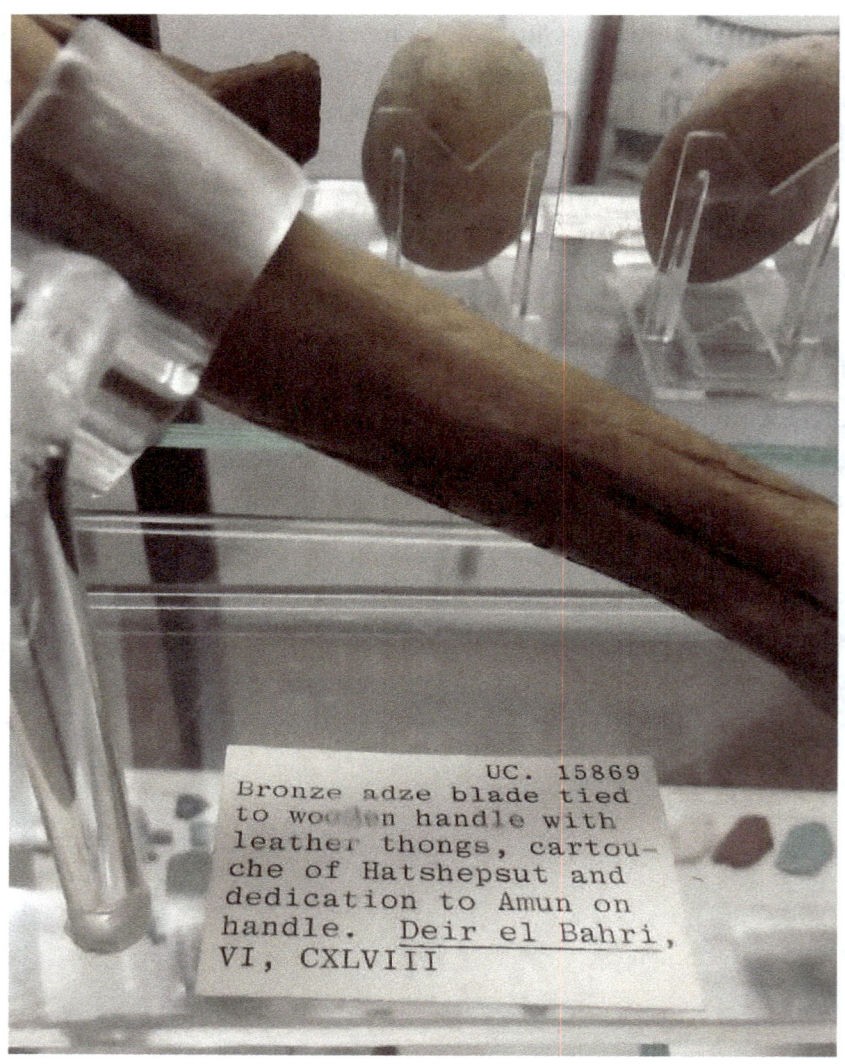

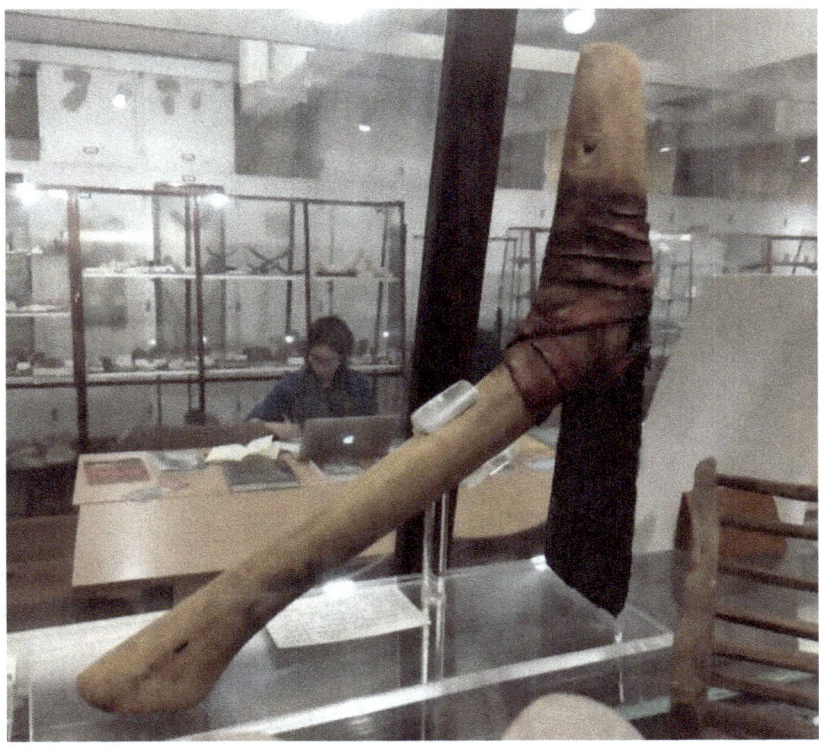

My initial assessment was that the damage was most likely caused by a sharp instrument swung downward, penetrating the skull. I believe that, if the blow didn't kill her instantly, her death would have been swift. Therefore, my guess would be murder, unless other evidence suggests otherwise.

In conclusion, I find it striking that out of the 7,000 bodies, only two skulls were kept, and the one I inspected in the museum showed signs of having been murdered. I suspect that the museum didn't want to display an elongated skull, which could explain why they chose this particular one—it was likely the only one deemed acceptable.

Before we even travelled to Malta, my research had already led me to search for potential weapons that could have caused such an injury. Given the circumstances surrounding this unresolved

THE GIZA PARK PROTOCOL HOW TO START THE SUN

genocide, where the entire population of Malta was massacred, totalling 7,000 casualties, I had already begun to consider various scenarios that might explain such a brutal event.

Today, I believe the most plausible explanation I can offer is this:

It's well documented that around 3000 BC, the consistent weather patterns began to change. Rainfall, which had been predictable, became less frequent, leading to crop failures and droughts. By 2500 BC, this period of water shortages was formally known as the "Dry Period," a devastating blow to Egyptian civilisation.

The prolonged drought caused extreme food shortages, which lasted for years, resulting in famine and starvation so severe that some records, including hieroglyphs, mention cannibalism. Additionally, many foreign powers, eager to seize Egypt's treasures, attacked weakened Egyptian communities, killing people, destroying temples, and looting their treasures. With the population already suffering, widespread unrest led to revolts against their leaders, creating a land of lawlessness and despair.

During this period, many Egyptians began to leave their homeland in search of better shelter and food. I suspect those who had access to boats on the Nile were the first to leave, as they could travel great distances relatively quickly.

If you visit the Giza Plateau today, you'll see five boat storage pits—places where the boats needed for daily operations at the Giza Industrial Park were kept when not in use. Inside the new GEM Museum, you can see a restored vessel that was removed from one of these pits, although the others were found empty. It's possible the missing boats were stolen and used by Egyptians to escape the dire conditions with their families and friends.

If I were in their position, as a USCG-licensed merchant marine captain, I would have likely used a similar escape method. Malta, a nearby country I was familiar with, would have been an ideal destination, especially if I had established a relationship with the people there.

A boat journey from the Nile, where it meets the Mediterranean Sea, would only take a few days to reach Malta. From the northern region of Libya, the journey would only take around 24 hours.

Now we must ask: why would they do this?

We know that the early Maltese people were megalithic builders and a community of farmers and fishermen. They had food, shelter, and a fleet of small fishing vessels—and they were geographically close. It's likely the Egyptians were familiar with these people, and if they were an elongated-skull race, they may have been taken to Malta to be removed from the Egyptian gene pool.

From my research, it's clear that throughout history, elongated-skull humans have been demonised, ostracised, and often murdered in an attempt to erase their existence.

This is precisely what happened to Pharaoh Akhenaten, who was branded a heretic and effectively erased from Egyptian history, along with his wife, Nefertiti. Both are believed to have had elongated heads, as evidenced by statues and hieroglyphs.

I want to note that my research keeps uncovering the presence of elongated skulls all over the world. Some, like those found in Peru, are naturally elongated, but most appear to have been artificially created through the use of devices that performed cranial deformation on newborn babies up to one year old.

For example, 260 elongated-skull corpses were discovered during excavations beneath the Quetzalcoatl Pyramid at Teotihuacan in Mexico. Additionally, numerous indigenous tribes

THE GIZA PARK PROTOCOL HOW TO START THE SUN

in the Americas, such as the Flathead Indians, used this technique on their children. One famous tribe, the Chinooks from Washington State, is well-known for their interactions with the explorers Lewis and Clark.

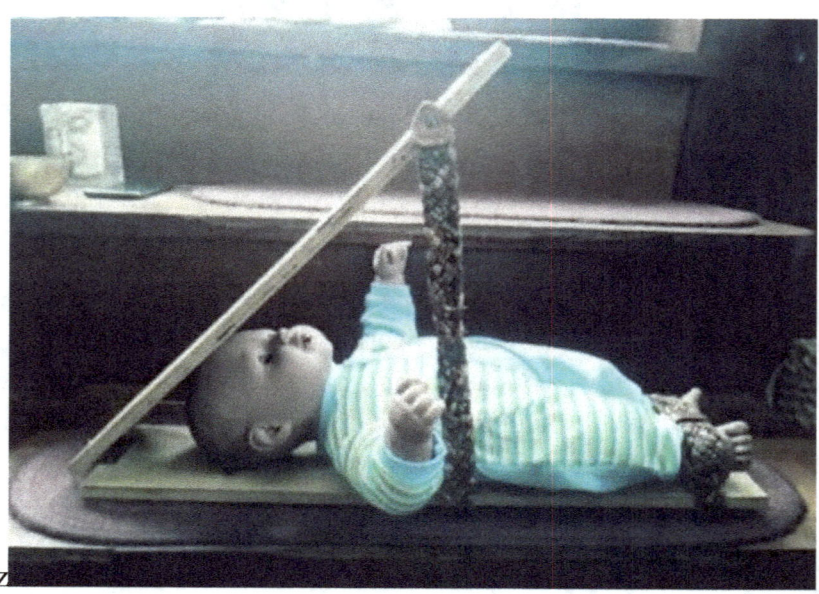

THE GIZA PARK PROTOCOL HOW TO START THE SUN

A Chinook woman with an infant in a cranial deformation device is pictured alongside the cranial deformation device I created for a lecture. This device demonstrates the procedure performed on newborns during the first 12 months of their lives.

An interesting side note about the Chinook Indian Nation today: they have been ostracised by the American government and have been rejected as a federally recognised Indigenous Native American Nation for the past 120 years. As a result, they do not receive the economic appropriations, medical care, or educational subsidies that other Native tribes in America do.

I wonder if this discrimination is linked to their ancient cranial deformation practices, which create elongated skulls.

From an outsider's perspective, having studied their history, it seems that there has been an international pattern of extermination and genocide, which appears to be ongoing, as though there is a concerted effort to eliminate this species from the Earth.

While discussing the names of countries, let's turn our attention to America. Another widely accepted but false narrative we have all been taught is that in 1492, a genocidal Spaniard by the name of Christopher Columbus was the first to "discover" America.

Columbus, a man responsible for horrific atrocities, left a trail of bloodshed, violence, and inhumane acts through the Caribbean islands, massacring thousands of Indigenous peoples. One particularly disturbing and perverse act he took pride in was playing games with children. He would line them up in two rows to compete in a "one-legged race."

Let me clarify what that "one-legged race" entailed. Columbus would take a sword and sever one leg from each child, then line them up to race against each other. The winner of the race was told that they would be killed quickly, without much pain, while the losers suffered a prolonged, agonising death.

THE GIZA PARK PROTOCOL HOW TO START THE SUN

It's hard to believe that we live in a country where numerous cities are named Columbus, and an annual national holiday is still dedicated to honouring this brutal man.

Alright, enough of the sickening details about Columbus.

The next topic will likely stir controversy. I must admit, I am far more interested in uncovering the truth than in appeasing any government official. The following explanation, accompanied by images, will shed light on why Columbus was chosen as the figurehead for the discovery of America—and how the true origins of America's name are much different than what we've been taught.

This will require a bit of explanation, so a brief lesson on Egyptian hieroglyphs is in order. Once I've finished, you'll see that the name "America" has Egyptian roots, contrary to the popular belief that it was named after the Italian explorer Amerigo Vespucci.

See the images below for reference:

#1 **PYRAMID:** The definition comes from the Greek word *pyramidos*, meaning "fire in the middle."

#2 **AMERICA:** The commonly accepted definition attributes the name to Amerigo Vespucci, an Italian explorer who followed Columbus. However, my research leads me to disagree with this origin. I am about to explain why I believe the name "America" is, in fact, Egyptian.

THE GIZA PARK PROTOCOL HOW TO START THE SUN

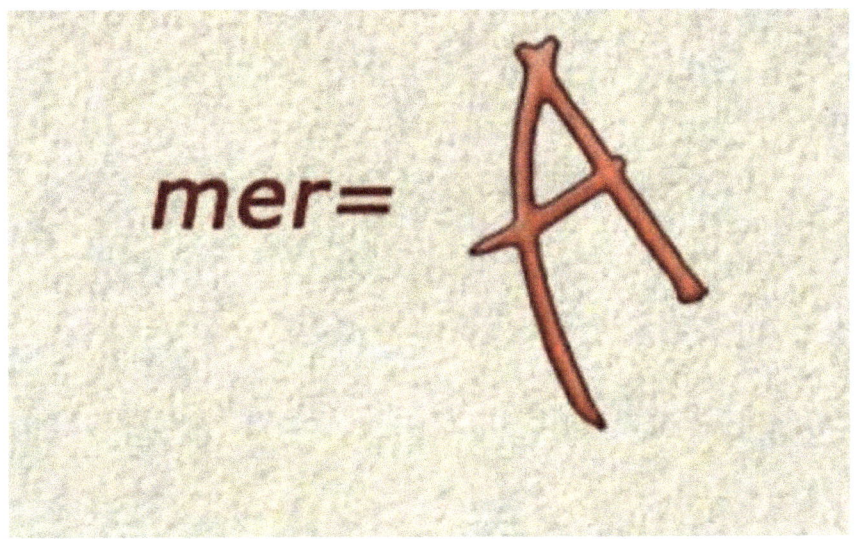

#3 Egyptian Hieroglyph "MER": Egyptian translation is "PYRAMID."

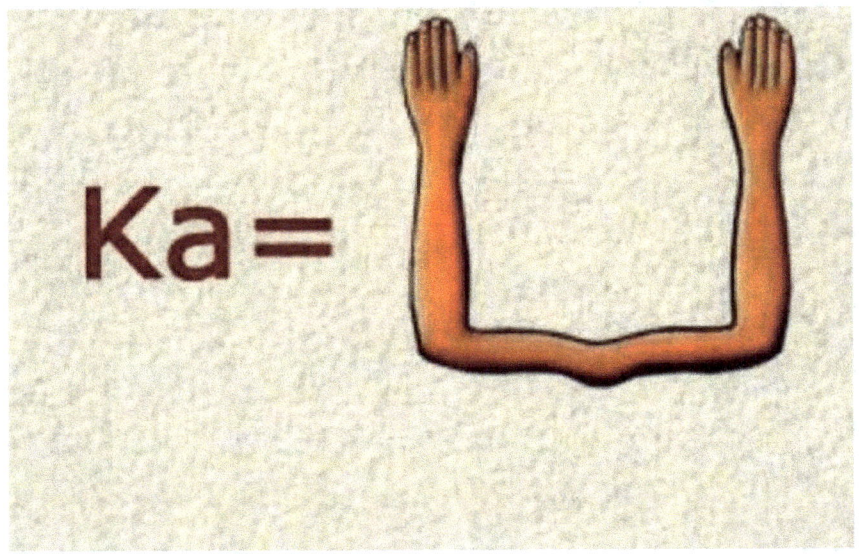

#4 Egyptian Hieroglyph "KA": Egyptian translation as "LIFE FORCE" or" SOUL." The open arms express the transfer of this "KA" energy.

Statue representing KA.

THE GIZA PARK PROTOCOL HOW TO START THE SUN

THE GIZA PARK PROTOCOL HOW TO START THE SUN

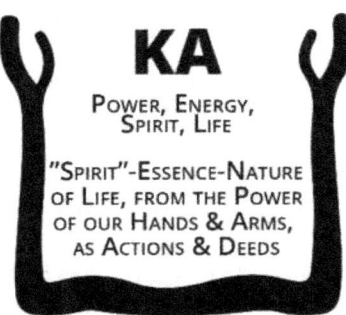

Ka/Kau - Bull/Cow. Power, strength, fertility. Power of action in the world to create from consciousness, self, soul, spirit, essence, etc. of who we are. Aspiring & rising up to higher consciousness.

KA

POWER, ENERGY, SPIRIT, LIFE

"SPIRIT"-ESSENCE-NATURE OF LIFE, FROM THE POWER OF OUR HANDS & ARMS, AS ACTIONS & DEEDS

Real spirituality is morality. Your spirit, soul, self & character is defined by what you do with what you know about reality and self/consciousness, your behavior relecting who you are.

Upright Arms, right-angle, Power/energy to Act with arms and hands. Our Moral "Spirit"-Essence-Nature determines our Quality of life. Character, Behavior, Actions, Deeds. Ka-racter, Ka-risma. Who we are, by our actions and deeds, determines our quality of life.

#5 OK, let's translate the word AMERICA into Egyptian hieroglyphs and proceed from there.

AMERICA

A MER I CA

MER CA

You can easily see where I am going with this discussion now, but wait until the end—I have a surprise!

My research of old-world maps revealed that AMERIKA has been spelled with a K longer in the past, not a C like today; however, they both have the same sound. I find it another enjoyable, clever distraction of truth once more.

It should be obvious now that AMERICA is Egyptian-influenced, the same as the name of ISRAEL. What a special relationship Israel, America, and Egypt have?

#6 Therefore, the Egyptian translation of AMERICA Symbolically could be explained in this way: MER = PYRAMID KA = LIFE FORCE or ENERGY

At this stage, America sounds precisely like the definition of "FIRE IN THE MIDDLE OF THE PYRAMID."

#7 Okay, here comes the real shocker I mentioned at the beginning of the lesson about the translation of AMERICA. We defined the MER and the KA; two letters are left over, exactly like with Israel.

AI

Those two letters are A & I. Did you get that AI?

What? Artificial Intelligence?

Now, I am inclined to express this once again: I believe that "America" is an Egyptian name, describing Pyramids, Energy, and

Artificial Intelligence. I don't know about you, but this is precisely why they don't want you to recognise the truth.

Remember, our current species, Homo sapiens, has been determined by science—information published by the Smithsonian Institute—to be the 21st evolutionary model of our species. Since I discovered this a decade ago, my theories have focused on the real possibility that our planet is an experimental environment designed to create new species. My research adds substantial collaborative evidence to reveal the truth, referencing their blatant arrogance while toying with our ignorance.

I grew up in the 1950s and 1960s in the small mountain town of Covington, Virginia, a papermill community. My small town was chosen as the most polluted city in America in the 1964-65 timeframe and was featured on the cover of an international magazine. The Jackson River, a stone's throw from my home for 18 years of education, was coal black with yellow suds floating on its surface. I never touched that contaminated, stinking water, nor ate any fish caught from it.

The air was always thick with the smell of rotten smoke and acidic ash, which covered our automobiles and properties most days. My grandfather, employed there for 47 years, had an automobile that he painted once a year. The Exxon service station, specialising in treating acid-ash-covered cars, used toilet bowl cleaners to remove the desiccated chemicals. Then, the car was hand-painted with a brush and given a thick coat of paint to protect the metal beneath.

Additionally, most food products grown and distributed to Americans today are genetically modified and lack nutrients. However, they contain substances that influence human hormone production. It is evident that with chemical manufacturers owning all the patented GMO seeds to grow the crops, America has become a chemical experiment—this, in addition to the introduction of fluoride chemicals into our water system during the 1950s.

THE GIZA PARK PROTOCOL HOW TO START THE SUN

There is no doubt in my mind that America is an AI experiment. If you disagree, prove me wrong!

Leaving the capital of the United States, I resumed my search, discovering the Chicago area—another location of immense interest in my findings. One of the main topics yet to be discussed in this text is energy grids, ley lines, vortexes, and nodal points. Not only is the Chicago area a powerful energetic nodal point, but a cave complex, shrouded in mystery, lies a few miles south of Chicago along the banks of the Embarras River. This river flows into the Illinois River, one of the numerous rivers that joined together to form the beginnings of the great Mississippi River, which flows to the Gulf of Mexico.

This makes it possible for a vessel of specific sizes to travel from the Atlantic Ocean into the Gulf of Mexico and up the Mississippi River, all the way to the Embarras River. Even today, despite shoaling over hundreds of years, you can still kayak the entire distance to Chicago.

As a retired USCG Merchant Marine Captain, I can guarantee that an Egyptian-style, seaworthy vessel, with an experienced crew, would have access to the location where the Burrows Caves are said to be. Like most other rivers formed by the melting of glaciers—some upwards of 2 miles tall as the termination of the Ice Age unfolded systematically—the Mississippi River is a product of those events, which I use to support my theories.

Let us examine some fascinating data. In 1982, Russell Burrows accidentally discovered Burrows Cave in Richland County, Illinois, USA. He claims the cave was a tomb filled with ancient relics, a storage site for the dead. The list of relics included gold statues, mummies, pagan idols, diamonds, coins, weapons, burial urns, and scrolls. These items reflected the influences of many ancient cultures, including Old World Egyptians, Romans, Hebrews, Sumerians, Greeks, and Phoenicians. The inventory was extensive, with thousands of clay and stone tablets carved to depict these civilisations. Numerous unconventional and ancient Egyptian artifacts were also found in the cave.

Many of these treasures have since disappeared, and many were sold to private collectors years ago. However, all archaeologists—including the US government—claim the entire discovery was a hoax, asserting that every artifact found is a fake. To convince the world and the public of the fraud, the US government declared the cave area unsafe and installed high-security fences, barbed wire, and "No Trespassing" signs.

I love a free and open country, but sadly, the US is far from it; censorship and human rights are waning rapidly in America.

While on this subject, another cave system, separate from my theories, adds to the collaboration of a second location in America with a similar discovery made earlier in 1909. The cave system, known as the "Underground Citadel," lies 2,000 feet above the Colorado River in the Grand Canyon. Soon after its discovery, the Smithsonian Institute in Washington, D.C. sent two archaeologists to investigate the newly discovered site in Arizona. Following their investigation, the two men were interviewed by a local newspaper, the Arizona Gazette, which then published an article about their findings.

According to Smithsonian archaeologists Professors Jordan and Kincaid, they discovered hundreds of Egyptian artifacts dating back to around 1250 BC, during Pharaoh Ramses' reign. They also found a vast network of tunnels with over 100 rooms carved out of the marble stone walls. Their assumption was that a community may have once lived there. As they ventured deeper into the cavernous tunnels, they found an entire collection of carved statues depicting Egyptian figures, and even discovered a giant Buddha statue carved from marble metamorphic rock.

THE GIZA PARK PROTOCOL HOW TO START THE SUN

I apologise if this sounds repetitive, but typically, all this data has been repressed since that time. The archaeologists have continued to push the same story: it's a hoax, fake, all a big lie. Is it?

Ironically, just like with Burrows Cave, the US government sealed off the area to "protect us."

Now, decades apart, two major discoveries in America are considered hoaxes. No artifacts were found, only fake relics. Again, the US government secured the discoveries to prevent further inspection.

If you contact the Smithsonian Institute today and ask them how many Egyptian artifacts have been found in America or the locations of such discoveries, I can tell you their response will be this: There have never been any Egyptian artifacts found in America. Have a nice day.

I want to stress that it is evident the Washington Monument was created as a symbolic Egyptian obelisk. Based on my discoveries, I believe the monument's connection to Egypt, particularly Giza, may explain why the US capital is located where it is. Furthermore, America has an original Egyptian pyramid featured on its one-dollar bill. I would also like to thank Robert Edward Grant for his mathematical research, which determined that the pyramid on the dollar bill is indeed Egyptian. He physically inspected the pyramids at the site at Abu Rawash, situated about 5 miles north of the Great Pyramid, on a large hill.

I have just listed two geographical locations in America where ancient Egyptian artifacts or treasures supposedly exist. If you prefer to say they have simply disappeared, both sites have been off-limits to the public, and absolutely no information has been provided by the US government—just silence.

THE GIZA PARK PROTOCOL HOW TO START THE SUN

I cannot imagine the logic behind these two unusual discoveries unless the Government is intent on withholding the truth from the public in this so-called "free" nation that supposedly treasures free speech and freedom from censorship. If true free speech existed, the US government would conduct a televised exploration of the world, showcasing how it prides itself on its freedom and free speech.

As I mentioned earlier, almost everything I was taught in the US school system about my homeland's history is a lie. Freedom in America—my latest term for the country—should be referred to as "Slavelandia." It's an illusion, and those who believe it exists here are delusional.

In a recent poll on global rankings of countries based on freedoms and rights, America was ranked 25th. So, there are 24 other nations offering more freedom than the United States, and I am actively seeking a place to live freely. Even worse, the latest polls rank the US 48th on the international list of nations by life expectancy.

More evidence is needed to support my theories and explain why I hold these beliefs. Persevering westward, curious about where my Osiris energy line would lead me next, I soon became fixated on the overwhelming number of anomalies that caught my attention in the Pacific Northwest states, beginning with Montana.

It's amusing to reflect on the fact that, even after more than a decade since my first encounter with Montana, no other state has captured my attention more than Montana, nor has any other state offered such valuable insights for my research. The results have turned out to be a profitable detour, providing strange and extraordinary information. Even today, I continue to spend much time conducting in-depth research there.

Through my work, I became close friends with Julie Ryder, gaining insights from her decades of hands-on, boots-on-the-ground exploration and research of ancient megaliths and dolmens. I affectionately call her the "Queen of the Montana Megaliths" due to the 111 incredible discoveries she has made.

Thanks to my relationship with Montana, I've added exceptional data to my inventory—"Tools in my Toolbox," as I call it. My fascination has led me to plan a trip with Heidi to visit Julie Ryder and her husband, Bill, for three weeks. During this time, I intend to explore some of my questions and witness Julie's extraordinary discoveries related to what is known as the "Flash Fossilization" of organic beings.

My interest in Montana also led me to discover, once again, the Flathead Indians—another elongated-skulled people, this time from the North American Indigenous Indian tribes. The US Government even named a National Forest system after them, calling it the Flathead National Forest.

What's great about America is that the government exiled all the Indian tribes from Montana, forcing them to leave the land. And so, the "free and humane" American government named their abandoned homeland after them. This situation feels eerily similar to how Israel named places after Egyptian gods.

America is, without a doubt, 100% about Egypt, through its connection to Israel, which has Egyptian roots.

The Flathead National Forest is located in Flathead County, south of Glacier Park. Spanning 3,758 square miles, it is named after the Flathead Native Americans from that region of Montana, including the Salish Tribe.

I also made a unique and highly unusual discovery in Montana, which forms a significant part of my theories. I've saved it for this part of our discussion, as it's a rare and extraordinary find that requires images to explain. (See below.)

THE GIZA PARK PROTOCOL HOW TO START THE SUN

This discovery is particularly meaningful to me because I'm a very visual person. I see many ideas in my mind, which I then draw. I have hundreds of original drawings.

By referencing various maps from different periods, I gain a broader perspective on identifying some of the locations that my unique energy line travels to.

With close attention to detail, I noticed the image of a face along the western border of Montana, next to Idaho. This image was captivating and ironic, especially in light of Julie Ryder's unconventional discoveries in the area. Upon further examination of this rare border face, I couldn't help but think of Abraham Lincoln.

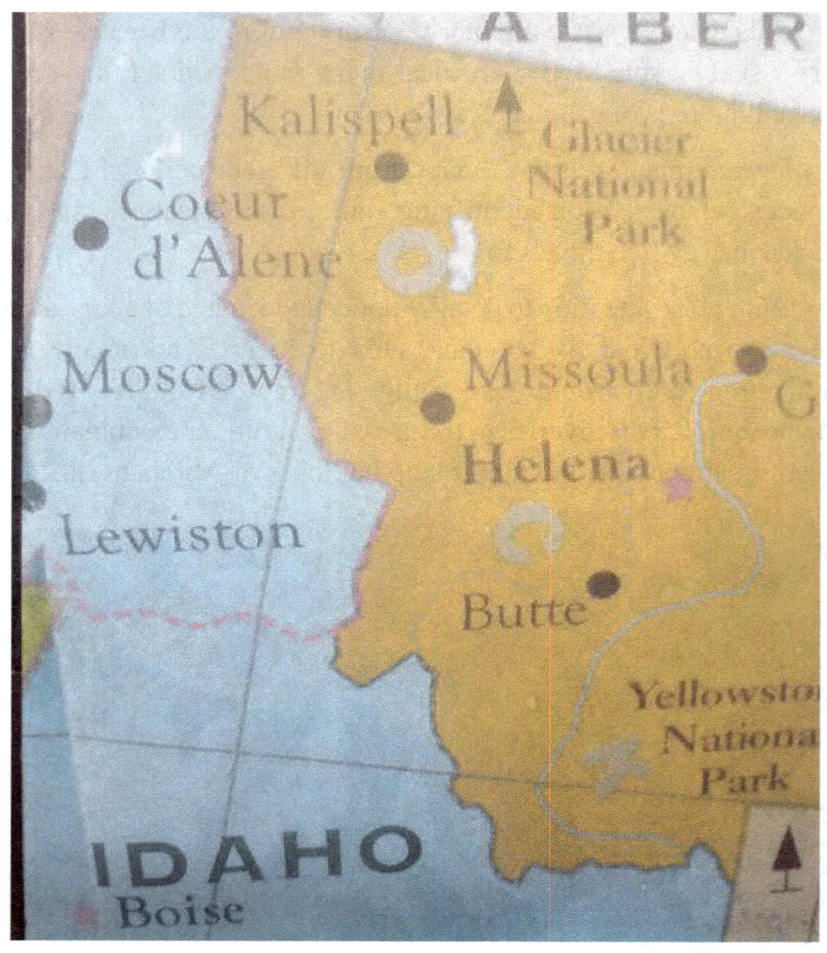

So, I decided to do a deep-dive forensics-style historical investigation into the 1800s and see how this freakish image on the west side of Montanna could happen. I thought, for a president's face to be used as the border of a state, give me a break.

Then I went to work, and this is what I found.

THE GIZA PARK PROTOCOL HOW TO START THE SUN

Montana is currently the fourth largest state in America—enormous and landlocked. Its topography is defined by the Great Hydraulic Continental Divide and the Rocky Mountains chain, which stretches from Alaska to Mexico.

I want to emphasise that geologists consider this Divide to be particularly special. It is a significant geological and continental hydrological divide. A watershed ridgeline functions as a drainage divide, with varying elevations that dictate how water flows across the land.

In 1864, when Abraham Lincoln was the presiding president, Montana was officially claimed as a US territory. He appointed the first governor of the territory just one year before his assassination. However, before this event, Lincoln had lofty ambitions and wanted a state named Lincoln State, as a memorial to himself.

Had the government granted his request, the new state would have encompassed what are now the northern parts of Idaho and the entire western portion of Montana. I want to focus particularly on the western part of Montana in this context.

Lincoln's historical popularity stemmed largely from his role in ending slavery in America, leading to the swift conclusion of the Civil War, which divided the country into two bitter halves. I find it difficult to view his decision as an act of fairness or humanitarianism, however.

Remarkably, his political resolve was rooted in a fascist approach to governance. To achieve his political goals, Lincoln could never be considered a kind-hearted human being; rather, he was a pragmatist willing to do whatever it took to implement his agendas. Unfortunately, his determination came at the expense of many innocent lives—particularly the indigenous peoples and the captured Confederate soldiers who suffered greatly under his actions.

A brief overview of Lincoln's turbulent and violent political career is in order. Before becoming a lawyer in Illinois, he was elected to the Illinois legislature in 1834. Throughout his career, he worked in various trades, acquiring many skills. He served as a boatman, store clerk, land surveyor, and militia soldier. He also served in the militia during the Black Hawk War, where his disdain and murderous attitude toward Indigenous peoples took root.

THE GIZA PARK PROTOCOL HOW TO START THE SUN

Lincoln was re-elected to the Illinois legislature four times, serving for 14 years, until 1848. In 1860, he was elected President of the United States. The famine and starvation that led to the displacement of Indigenous peoples contributed to a violent uprising by Sioux tribes against white settlers in what became known as the Dakota War of 1862, or Little Crow's War.

The war lasted only five weeks, resulting in the deaths of hundreds of settlers and the displacement of thousands more. Afterward, the entire Dakota Nation was forced from the territory. Of the 303 Dakota Indians captured, 39 were sentenced to death. Lincoln, wanting to make a grand spectacle of his stance against the Native Americans, oversaw a brutal execution.

On December 26th, 1862—the day after Christmas—Lincoln gave the American people a horrific Christmas "gift." This is a striking example of what is now often referred to as "Making America Great Again." Even the Guinness Book of World Records recognises Lincoln for his role in planning one of the largest public executions in American history.

This execution was not just any ordinary affair; it was a massive public event, with thousands of white settlers gathering to witness the hangings. Lincoln designed gallows that could accommodate 38 men at once. To this day, this remains the largest mass execution in America's history.

TION OF THE THIRTY-EIGHT SIOUX I
AT MANKATO MINNESOTA DECEMBER 26, 1862.

To clarify, on American soil, during WWII, the US dropped two nuclear bombs on civilian populations in Japan, massacring families. The most reliable estimates of the death toll vary, but the consensus places it between 110,000 and 210,000 lives lost.

In contrast, Abe Lincoln's desire for a state of his own was ultimately denied, which left him deeply disappointed. He had even gone so far as to design a state flag and plan a government cabinet for his envisioned state. His dream was shattered after years of planning, but Lincoln's pragmatism and resourcefulness allowed him to partially achieve his goal.

This discovery is significant because it adds yet another piece of corroborating evidence to my extensive list of findings and further supports my theories.

Before Montana could officially become a state, its borders had to be surveyed and defined. A land surveying company was contracted to determine the boundaries, but President Abraham

THE GIZA PARK PROTOCOL HOW TO START THE SUN

Lincoln insisted on overseeing the entire process. With his expert training as a land surveyor, Lincoln demanded to be the sole authority in charge of the surveyors. Over time, the surveyors produced five different border designs for the new state of Montana—rather than the state of Lincoln.

As you can see from the images below, this aspect of my research has fascinating twists and turns. I've created overlays to highlight the unmistakable similarities, showing that Lincoln was, without a doubt, quite a savvy strategist.

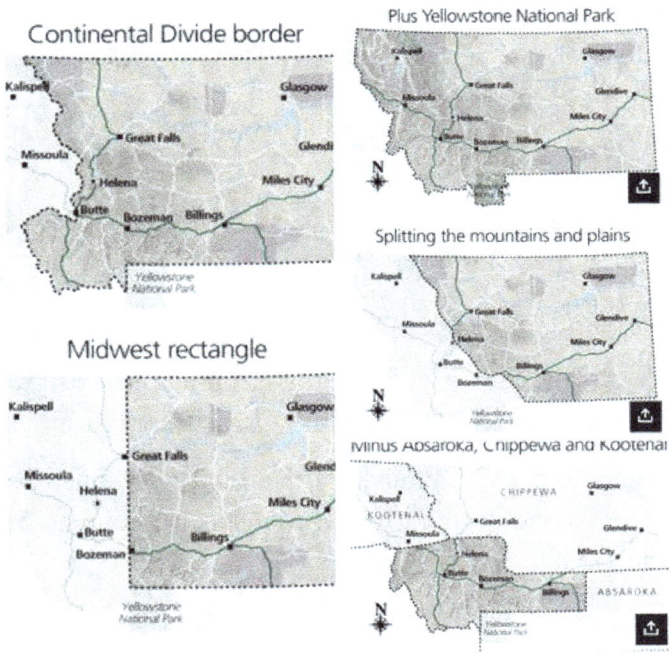

Working directly with Lincoln, the one design chosen by him out of 5 was chosen and still exists today.

See below:

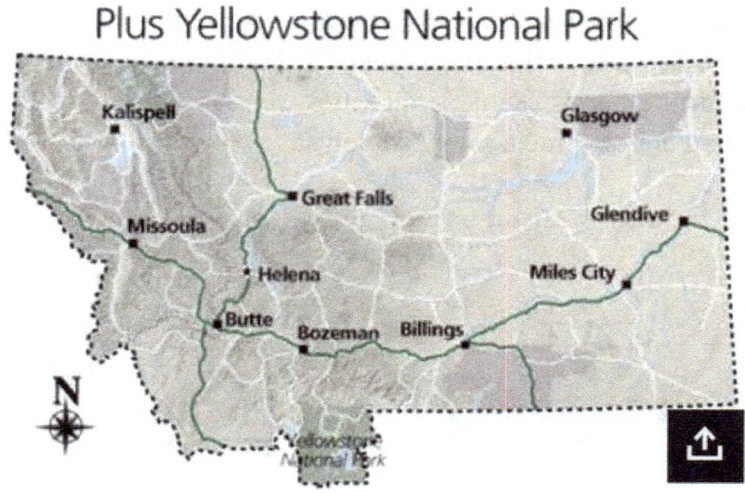

The images tell a story! Montana has much more to reveal, so let's move westward and focus on my Osiris Energy Line Pathway.

There is more to Montana, and I promise to revisit it soon to complete the second chapter, which connects to Montana, Abraham Lincoln, and much more.

I soared over Idaho's short land distance without any apparent reason to stop at a particular location. Eventually, I landed in Washington State like a meteor impacting the Cascades Volcanoes region—a region directly linked to the "Pacific Ring of Fire."

At the time, I didn't realise its significance. My discoveries here would educate me on how global systems were designed—how every tiny detail was fabricated or manipulated, incorporating every available resource to create the desired effect.

THE GIZA PARK PROTOCOL HOW TO START THE SUN

I believe the information I am sharing has been available for a very long time—over 32,000 years, perhaps even longer. I feel this knowledge can be unlocked through an understanding of Pyramid Systems.

When I first became interested in the Great Pyramid and its operation in Egypt, I would call Egypt and speak with those overseeing the systems on the Giza Plateau. One of the departments responsible was the Cairo Water Department. The learning curve was steep and challenging, but I eventually established good contact with someone who provided me with historical information regarding my curiosities.

One of my inquiries stemmed from my interest in the mysterious or legendary Hall of Records. I had heard claims that all the tunnels beneath Giza were once filled with water and that archaeologists and researchers had, at one point, swum through them.

The gentleman I spoke with confirmed this. He explained that when they set up pumps to remove the water from the tunnels, it took four years of continuous, 24-hour-a-day pumping from multiple pumps to fully drain them.

This is why I argue that the ongoing search for the Hall of Records—a supposed physical location containing a vast library of secret knowledge—is misguided. If the Hall of Records were indeed a library filled with books, papyrus, or any material other than clay tablets or stone inscriptions, it would have been completely destroyed by water.

Thus, the search itself may be a distraction from the truth. At this point, ignorance is no longer an accident—it is a choice, resulting from failed research conducted in the wrong manner.

I admire Edgar Cayce. I had the pleasure of enjoying two Christmas dinners with friends who now own his original residence—the very house where he passed away. I also find many of his extraordinary readings to be both fascinating and valuable.

However, Edgar's style of speech and phrasing can sometimes leave one uncertain, guessing, and even confused.

When I followed his directions regarding the Hall of Records, he was specific: it lay beneath the Sphinx's left paw, between that location and the Nile. This led me to Osiris's Tomb and the surrounding area. It was here that I discovered Osiris's Tomb was directly connected to the Sphinx via a tunnel.

Toward the end of this treatise, I have included a chapter titled **"The Greatest Discovery at the Giza Plateau: Osiris's Tomb."**

Pursuing the Hall of Records beneath the Giza Plateau is a complete waste of time—unless one approaches the search from a different perspective. That perspective worked perfectly for me. My approach involved identifying operational systems, employing reverse engineering methods, and understanding how and why these systems functioned.

The results, which I consider both significant and fruitful, revealed knowledge that could have been found in an instruction manual—but wasn't.

I can only speak from my own experience and perspectives, which have evolved and shifted many times over the last 15 years of passionate forensic investigation and research.

Once you have absorbed the discoveries I share, I hope you will reconsider the version of history we have been told—a version that is more **fairy tale** than truth.

The final chapter of this text is something to look forward to. It unveils a new interpretation of one of the most famous Egyptian hieroglyphs: **the Papyrus of Ani, Plate #2.** This interpretation has the potential to redefine the meaning of thousands of hieroglyphs and may even necessitate a rewritten, updated version of the **"Egyptian Book of the Dead."**

Ultimately, you must decide what you take away from this **greatest story ever told**—a story of our past from multiple perspectives.

In this case, I believe the goal was to **end the Ice Age** and create a human-compatible environment. The intent was to populate the Earth with humans in a uniform manner, providing a planetary environment suitable for worldwide relocation—and they achieved this brilliantly.

Antarctica was the only landmass left uninhabited due to its extreme environment. Otherwise, we would have settled there as well.

From here, my journey continued across the state named after America's first president, **George Washington.** Today, his image appears on the US one-dollar bill—where an **Egyptian pyramid** is prominently featured on the reverse side.

An interesting coincidence.

You will notice that both **George Washington** and **Abraham Lincoln** frequently appear alongside **Egyptian symbolism—often in grand, monumental ways.**

So far, George Washington has the **Egyptian-style Washington Monument** and an **Egyptian pyramid** on the reverse of the one-dollar bill.

THE GIZA PARK PROTOCOL HOW TO START THE SUN

See images below:

Washington State, which borders the Pacific Ocean, was named after **George Washington**, the first President of the United States. In contrast, the capital of the United States, **Washington, D.C.**, is located on the East Coast, near the Atlantic Ocean.

I have made a remarkable number of profound discoveries in Washington State—findings that I believe will influence how scientists and researchers study **Pyramid Systems** and **Volcano Systems** in the future.

As I write these reflections from Mexico, **Heidi and I** are at the end of a four-month journey we call our **"Sea to Shining Sea Tour."**

This adventure began in **Washington, D.C.**, and took us across the United States of **Mer Ca** to the Pacific Ocean.

THE GIZA PARK PROTOCOL HOW TO START THE SUN

Along the way, we trekked to Ohio to explore the famous Serpent Mound, where I introduced the first image of my modern version of my discoveries of the Famous Greek God Hermes Armillary.

THE GIZA PARK PROTOCOL HOW TO START THE SUN

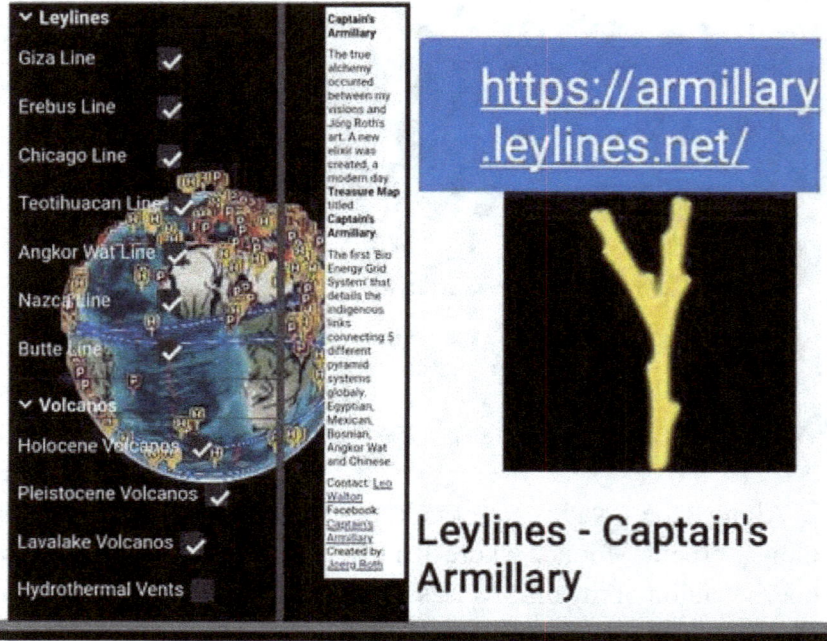

I call this interactive app creation the **"Captain's Armillary."**

We then attended the **5th International Scientific Pyramid Conference: Pyramid Secrets** in Chicago, where we had the privilege of meeting one of my greatest influences—**Chris Dunn**, author of *The Giza Power Plant*.

I also deeply admire **Dr. Semir Osmanagić** for his groundbreaking discoveries at Bosnia's **Pyramid of the Sun**. He was in attendance, and Heidi and I are excited to join his tour during the **Summer Solstice in 2024**.

It was also a thrill to finally meet my long-time friend **Julie Ryder**, the **Queen of the Montana Megaliths**, along with her talented husband, **Bill**.

Our research and discoveries continue to expand, adding to this extensive project, which explores lost history from new perspectives.

Next, we traveled to **Palm Springs, California**, to attend the **CPAK XII Conference** on *Precession and Ancient Knowledge,* hosted by the **Binary Research Institute** and its founder, **Walter Cruttenden**. It was an outstanding event, headlined by the renowned **Graham Hancock**.

There, we met the well-known **Patricia Awyan**, celebrated for her Egyptian research and tours. We also had the pleasure of reconnecting with **Chris Dunn**, who eagerly awaits the release of his new book.

Additionally, we met the **new "Water Wizard"** from New Zealand, **Veda Austin**, who has elevated the study of water consciousness to a new level. Her research captures frozen "water memory images" that provide fascinating insights.

Now, let's dive into my discoveries in **Washington State**, beginning with the purpose of Heidi and my **"Sea to Shining Sea Tour."**

My research in **Egypt** led me to an obsession with the **Great Pyramid**, and this fascination has since expanded globally. As a result, I have connected with countless researchers worldwide who share similar interests.

That's exactly what happened in **Washington State**. I collaborated with someone who believed they had discovered a **pyramid** there. Over time, I became convinced it was worth investigating in person. Unfortunately, due to a sudden illness, the collaboration fell through.

Rather than letting our planned trip go to waste, we continued our journey, traveling near that area on our way to **California**. We stopped in **Portland, Oregon**, near the **Columbia River**, to study some of the volcanoes and geological formations connected to my research.

THE GIZA PARK PROTOCOL HOW TO START THE SUN

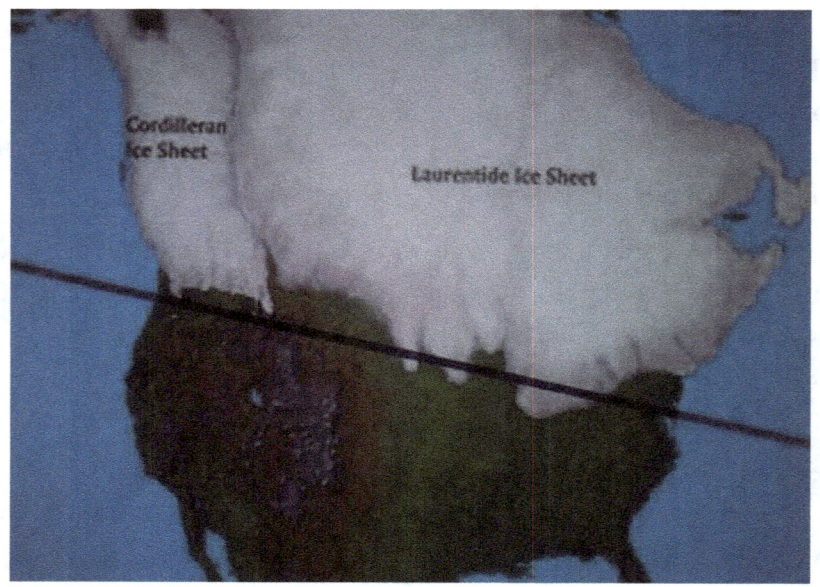

I discovered that my **Energy Line** intersected a region known as the **Cascades Volcanic Arc**—a 700-mile-long subduction zone stretching from **British Columbia, Canada**, to **Northern California**. This region contains numerous active volcanoes, revealing a deeper connection to my findings.

Below are images of **North America's Ice Sheet** during the **Ice Age** and the **Osiris Energy Line Path**. I have drawn on the images to give you a clearer idea of the overall **"Sea to Shining Sea"** pathway.

What I first observed when I first made this discovery over 15 years ago was I determined and found it interesting that the Ice Sheet never formed south of this Osiris Energy Stream.

It reveals how a perfectly designed and executed agenda is based on a genius plan to melt the Pleistocene Ice Age, which had gripped the Earth and held it hostage for 2.4 -3 million years.

THE GIZA PARK PROTOCOL HOW TO START THE SUN

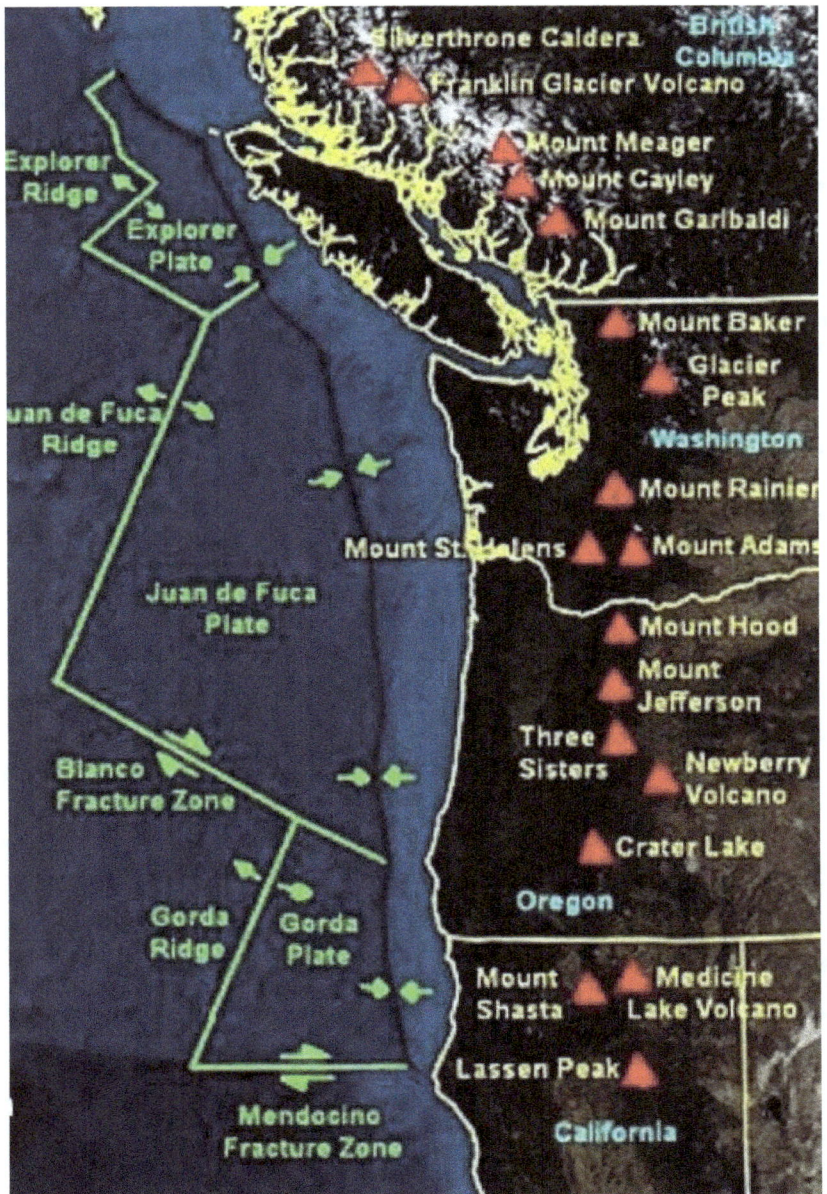

My discovery, the Osiris Energy Line, struck head-on into what I thought was a mountain named Lone Butte, Washington State.

However, my research proved otherwise. Lone Butte is called a flattop or Tabletop Mountain but is a volcano.

My curiosity deepened as I began asking more questions. **Lone Butte** didn't resemble any volcano I had ever seen, whether in person, images, or videos.

I soon discovered that **Lone Butte** is an exceptionally rare type of volcano—one of only three in the **United States** and the only one in **Washington State**.

After digging into its history, I learned that **Lone Butte** has its distinctive, flat-topped appearance because it is a **subglacial volcano**, specifically a **Tuya-type volcano**.

A **subglacial volcano** can only form when it erupts **beneath a massive ice sheet**—in this case, during the **Ice Age**, when the region was buried under a **two-mile-thick** glacier.

During the eruption, the immense weight and thickness of the ice **prevented the magma from rising freely**, forcing the volcano to take on its unusual **flat shape** instead of the classic conical form.

A Stunning Alignment of Subglacial Volcanoes

Intrigued, I set out to locate the other two **subglacial volcanoes** in the **United States**—and what I found was astonishing.

I didn't have to search far! The other two, **Hayrick Butte** and **Hogg Rock**, are located **side by side** just a few miles **north of Mount Washington**, in **Oregon**.

Even more fascinating, all three volcanoes are **perfectly aligned** in a straight line, running **south from Lone Butte**, just across the **Columbia River** near the **Oregon-Washington** border.

Here's the breakdown of their distances:

Mt. Hood is **72 miles** from **Mt. Washington**.

Mt. Washington is **118 miles** from **Lone Butte**.

THE GIZA PARK PROTOCOL HOW TO START THE SUN

The **perfect linear alignment** of these subglacial volcanoes ignited my curiosity further. I had to dig deeper—where else could I find these rare formations?

As I expanded my research into the **northwestern regions of North America**, including **British Columbia** and **Alaska**, I came to appreciate the sheer **complexity and ruggedness** of these environments.

This region was **shaped by extreme forces**—massive volcanic eruptions and catastrophic floods from glacial meltwater, which **reshaped entire landscapes**.

My research soon bore fruit. **Straight north of Seattle, Washington**, across the **Canadian border** into **British Columbia**, I found the first of over **40 Tuya subglacial volcanoes**!

I believe that all these subglacial volcanoes **likely erupted simultaneously**.

Take another look at **Lone Butte**—notice that it appears different from the others. There is a **good reason** for this distinction.

In **shape**, it resembles a **shield volcano** more than the typical **flat-topped tuya**.

Lone Butte was likely the **first** of these volcanoes to erupt during the **Ice Age**, beginning its formation beneath a **two-mile-thick glacier**.

Over time, **erosion beneath the ice** created an **open cavity**, allowing Lone Butte more **room to grow**. As the **ice melted from below**, the **distance between the volcano and the ice sheet increased**, enabling it to develop a **shield-like structure** rather than a **tabletop shape**.

The Mystery of Ice-Age Volcanic Eruptions

For these **frozen volcanoes** to erupt **after millions of years beneath ice**, some **extraordinary energy source** must have been involved.

These eruptions **melted massive amounts of ice from below**, creating **subglacial rivers** that disturbed the stability of the **Ice Age environment**.

To me, this reflects **pure brilliance and precision**—a meticulously executed process.

The Debate Over the Ice Age's End

When reconstructing events, our goal is to **gather as much data as possible**, ensuring consistency with known facts while **connecting the dots** based on **scientific evidence** across multiple disciplines.

Many scientists today believe the **Ice Age ended in as little as 4,000 years** due to **meteor impacts**. They argue that these impacts **heated the atmosphere**, triggering a **rapid rise in global temperature**, which ultimately **melted the two-mile-thick ice sheet**.

THE GIZA PARK PROTOCOL HOW TO START THE SUN

While I acknowledge the reality of **meteor impacts**, I do not fully accept this **mainstream hypothesis**. To me, this explanation **feels like a convenient theory**, created to fill a **gap in understanding** rather than based on a **realistic mechanism**.

A Different Perspective

First, I disagree with the proposed **short time frame** for the **Ice Age meltdown**.

While I **do** believe the ice melted **faster** than conventional models suggest, the amount of **ambient heat** retained by the Earth was **not sufficient** to explain the rapid ice loss.

Faced with this inconsistency, scientists have **grasped for alternative explanations** to justify their calculations. However, I believe I have identified a theory that **fits better**.

You decide.

Introducing My Theories

We have now **traveled across Africa, the Atlantic Ocean, and the United States**, exploring new data points. Now, my **theories begin**:

I have compiled a **detailed catalog** of **major Earth-changing events**—including **floods, volcanic eruptions, and theories on the Ice Age's termination**.

This timeline starts with the **Age of Scorpio (14,500 BC)**.

Another key period occurs near the **end of the Age of Taurus (2,500 BC)**—a time known as **Egypt's "Dry Season"**, marking a global shift in population.

I also introduce my **analysis of "The Energy Paradox"**, a scientific mystery concerning the **speed at which the Ice Age ended**.

THE GIZA PARK PROTOCOL HOW TO START THE SUN

Rethinking the Younger Dryas Event

My take on the **Younger Dryas impact hypothesis** differs significantly from the mainstream view that a **meteor struck Greenland**, causing global cooling and ice sheet collapse.

I have **no doubt** that impacts occurred. However, I have **first-hand experience** in one of the **largest known impact zones in North American history**.

Through **field research**, I have:

Physically investigated impact sites.

Measured energy grids on both **land and water**.

Analyzed scientific field data on the **immediate and long-term effects** of the event.

One of the most **notable results** of such an impact is the **Chesapeake Bay**, formed when the **Susquehanna River** flooded into a **300-foot-deep crater** left by a massive meteor strike.

Energy Release from Impact Events

The **violent impact** of a **50,000-mph meteor** caused extreme geological reactions, including:

Shock metamorphism, which altered rock structures.

Kinetic energy displacement, creating an **impact excavation zone**.

Compression-induced lightning discharges, which can form **Fulgurites** (fused glass tubes) in sandy environments.

I spent **three years** studying and **collecting Fulgurites** that were dredged from shallow waters and pumped onto nearby beaches during reclamation projects. These formations provide **direct evidence** of the energy released during impact events.

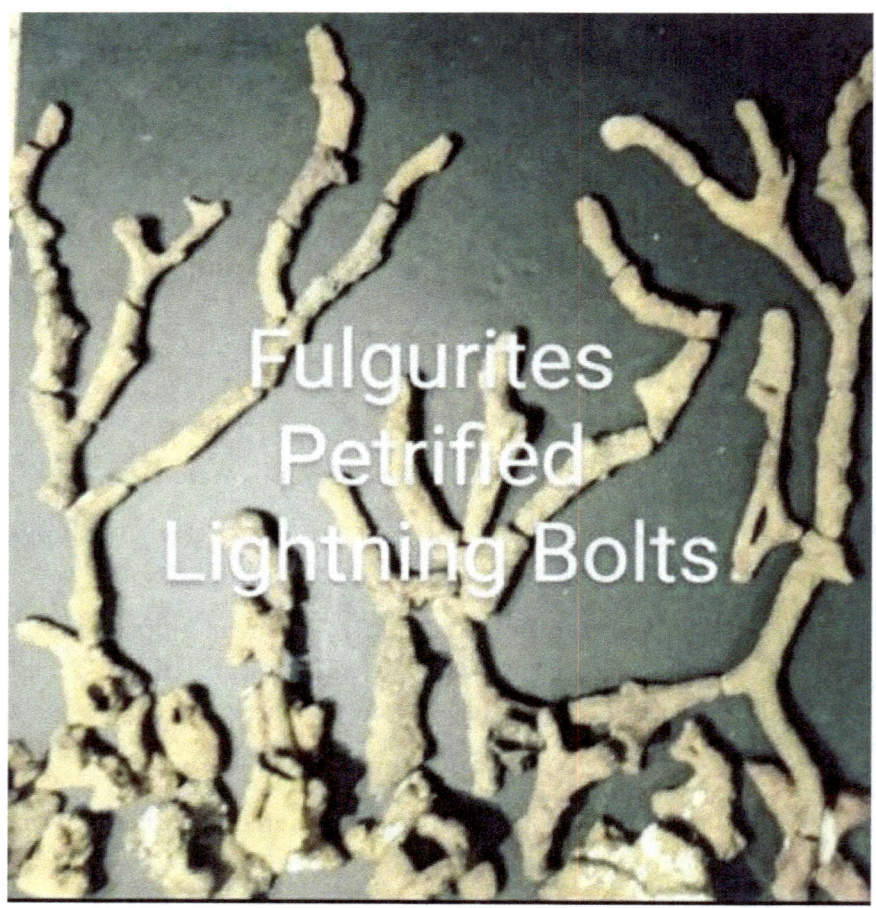

My research on the long-term effects of this event on geography and climate found consensus within the scientific community. There were no catastrophic consequences—only short-term influences on the atmospheric environment. Scientists concluded that the Chesapeake Bolide had an insignificant impact on the planet's weather and climate and did not cause the Ice Age to melt.

I am convinced that some meteor impacts occur due to an increase in the Earth's core's cyclic rotational speed. This acceleration extends the Earth's magnetic field further into space, creating a larger magnetic net that attracts and captures incoming meteors, particularly those composed of iron or other magnetic minerals.

I have always excelled at brainstorming—considering every possible explanation for a phenomenon, laying them all out, and selecting the one that best fits the situation. Brainstorming is a method that grants us the freedom to visualise our world differently, both naturally and logically.

Having made this statement, I wish to inform everyone embarking on this journey that the "truth" is not always pleasant—and in today's world, it is in increasingly short supply.

Furthermore, weather and climate modification systems are expanding every year. I include HAARP Ionospheric Heater Systems, cloud seeding (both ground-based and airborne), and CERN-type projects in my research, as they may influence volcanic activity and earthquakes—both of which have direct effects on weather and climate.

Through my research, I have determined with certainty that the Earth experiences "mood swings" in terms of weather and climate. It follows a natural cyclical pattern.

For most of my adult life, I have studied ancient history, philosophies, religions, and civilisations from around the world. I have read the Sumerian texts and understand the Earth's ancient black, barren state before it had an atmosphere. I recognise that Earth was once a single landmass, and I acknowledge the possibility that our Moon is both artificial and hollow.

I have been fortunate enough to understand Vedic sciences and possess a talent for deciphering symbolism. By profession, I am a systems analyst.

I believe that both the Earth and humanity are subjects of experimentation by a more advanced intelligence. I also theorise that, in ancient times, systems were constructed on Earth to shape and facilitate the environment—perhaps as part of this experiment.

All major catastrophes, I suspect, are either meticulously planned, externally influenced, or foreseen in advance. I believe that Pangea was deliberately divided into continents through the construction of structures designed to control landmass separation, volcanic activity, and earthquakes—perhaps even through the use of pyramid systems.

I am about to reveal to the world a network of pyramid systems—one in Egypt, one in Bosnia, one in Mexico, one in Cambodia, and one in China—that operated in conjunction to end the Ice Age.

THE GIZA PARK PROTOCOL HOW TO START THE SUN

This was done to aid and preserve the remaining human survivors, allowing them to move across the Earth and multiply, ultimately creating a new, large population for experimentation—currently standing at 8.1 billion.

Therefore, the notion that pyramids were merely constructed as tombs for pharaohs is a laughable misconception.

I also believe that modern civilisation has utilised many of Nikola Tesla's inventions and ideas, modifying the weather and climate on a global scale for years. Tesla was correct—within 100 years, his work would blanket the Earth.

Now, having travelled most of the way across America, we have uncovered some fascinating insights.

From our vantage point at the Lone Butte Tuya Volcano in Washington State, USA, we observe that the Egyptian Osiris Energy Path connects to a rare subglacial volcano—one of only three in America. This marks the second significant location on this journey, the first being our arrival atop the Washington Monument in the capital of the United States, Washington, D.C., named after the first U.S. President, George Washington.

I do not believe this connection was ever widely known and is likely understood by only a select few.

Why?

To avoid confusion, I want to complete the full pathway of the Osiris Energy Line to the Pacific Coastline and then return to Montana to incorporate the missing components. This will clarify and explain how and why these systems function. However, before diving into the specifics, I must provide a global overview.

From the Lone Butte perspective, we find ourselves in one of North America's most geologically active volcanic regions. The Cascade Range contains five major volcanoes and three subglacial volcanoes, all of which erupted during the Ice Age.

THE GIZA PARK PROTOCOL HOW TO START THE SUN

Lone Butte itself is located less than 20 miles northeast of Mount Adams. To the northwest, less than 20 miles away, stands Mount St. Helens. Measuring 25 miles to the west, we encounter another minor volcano, Tum Tum Mountain.

Using Google Earth, I have created a series of intriguing geographic images, highlighting key alignments and demonstrating the use of sacred mathematics as a blueprint for construction.

In the image, Mount St. Helens is positioned at the top, Lone Butte at the bottom right, and Tum Tum Mountain at the bottom left.

The formation bears a striking resemblance to either a volcano or a pyramid. To me, this is valuable evidence suggesting a deliberate design behind pyramid systems and their close relationship with volcanoes.

THE GIZA PARK PROTOCOL HOW TO START THE SUN

First, we must recognise—based on evidence—that pyramids and volcanoes are interconnected systems. Once we understand this connection, we must accept its implications.

Even if we disregard the many other volcanoes in the vicinity, this location alone presents a significant energetic trilogy, with Mount St. Helens as a key component. It remains an active volcano, with its last major eruption occurring in 1980. I remember that event well and personally know many people who lived there at the time.

I share this with you because you will encounter this pyramid-like formation repeatedly at different locations worldwide. Is it a message? I believe so—but I leave it to you to decide.

Remember our earlier discussions on the Egyptian and Greek meanings of the word "pyramid." I am particularly drawn to the definition **"fire in the middle."** Let me explain.

Beneath this image, I reflect on the Osiris Energy Line and its connection to this location.

As we continue our journey toward the Pacific Coastline, I will share with you the many treasures I have uncovered along the way.

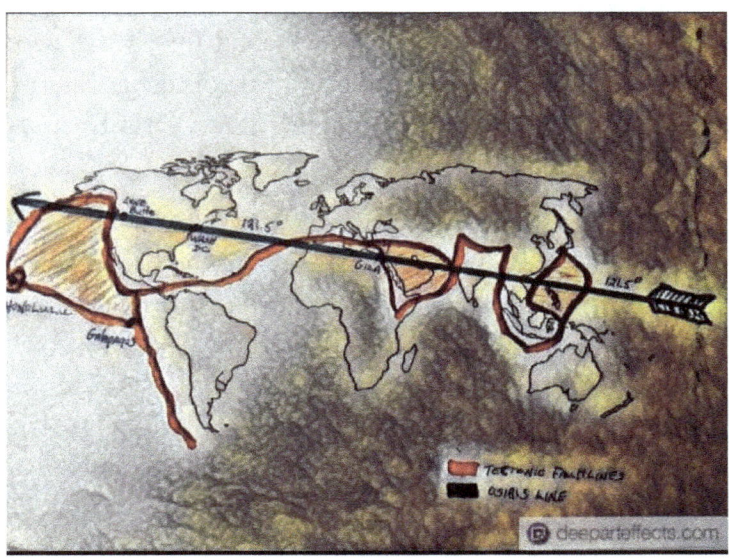

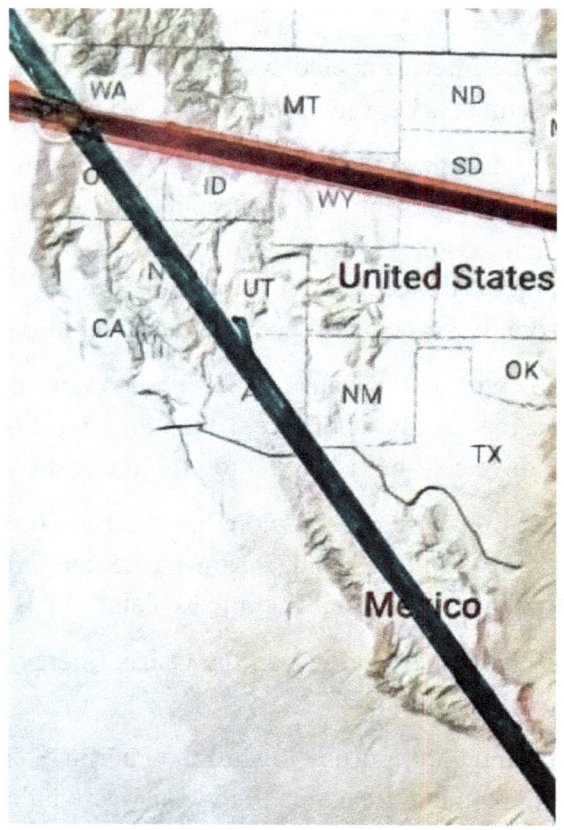

As you can easily see, I added a second line, which you will also learn about. I said it now because the Teotituican Energy Line Path Intersects is at the exact location of the Osiris Energy Line, which is at Lone Butte.

Once I made this second discovery from Mexico, I realized I was becoming very successful with my research and accurately interpreting the messages I received from the information provided by the Hall of Records in Akosha.

THE GIZA PARK PROTOCOL HOW TO START THE SUN

This is one example of my research, **"Fire in the Middle"**—a concept that utilises the Great Pyramids as a **Reaction Chamber.** The pyramid-shaped formation in the Cascade Range consists of three volcanoes: Mount St. Helens, the Tuya Subglacial Volcano, Lone Butte, and Tum Tum Mountain.

As we stand here at Lone Butte, we again turn our focus to the coastline, following this **Treasure Map** provided by the **Hall of Records,** which directs us with remarkable precision.

During my investigation of this region, I uncovered some unexpected distance measurements in miles—alignments that suggest deliberate design.

Following a **heading of 243 degrees,** we travel precisely **54 miles** to reach the heart of **Longview, Washington**—a city shaped like a pyramid and named after the successful businessman **Robert A. Long.** The Osiris Energy Line lands **exactly** at **R. A. Long High School.**

THE GIZA PARK PROTOCOL HOW TO START THE SUN

If that isn't fascinating enough, another precise measurement of **54 miles from Lone Butte** leads us to **Exodus Church**—a remarkable physical link to **Egypt, pyramids, and Israel.**

The connection to **Exodus** is particularly intriguing. In ancient history, Exodus represents the departure of the Israelites from Egypt—a direct link between these locations and the historical significance of pyramids.

Furthermore, the number **54** holds sacred geometric significance. **5 + 4 = 9**, a number often associated with the **Sun** or **God.** The **243-degree heading** also reduces to **9 (2 + 4 + 3 = 9)**, reinforcing this pattern.

We must consider the significance of numbers and numerology. For thousands of years, numbers have been used as **codes**—hidden keys to unlocking ancient mysteries.

See image:

THE GIZA PARK PROTOCOL HOW TO START THE SUN

Again, I turn on my navigation and head down the navigatable Columbia River, which borders Longview.

Continuing our heading of 243 degrees, we travel an additional 54 miles and end up at the mouth of the Columbia River, where it enters the Pacific Ocean.

This inlet is bordered by Long Beach and named after the same person. The total distance from Lone Butte is exceptional and one of the most sacred numbers: 108 miles, again 9.

See images:

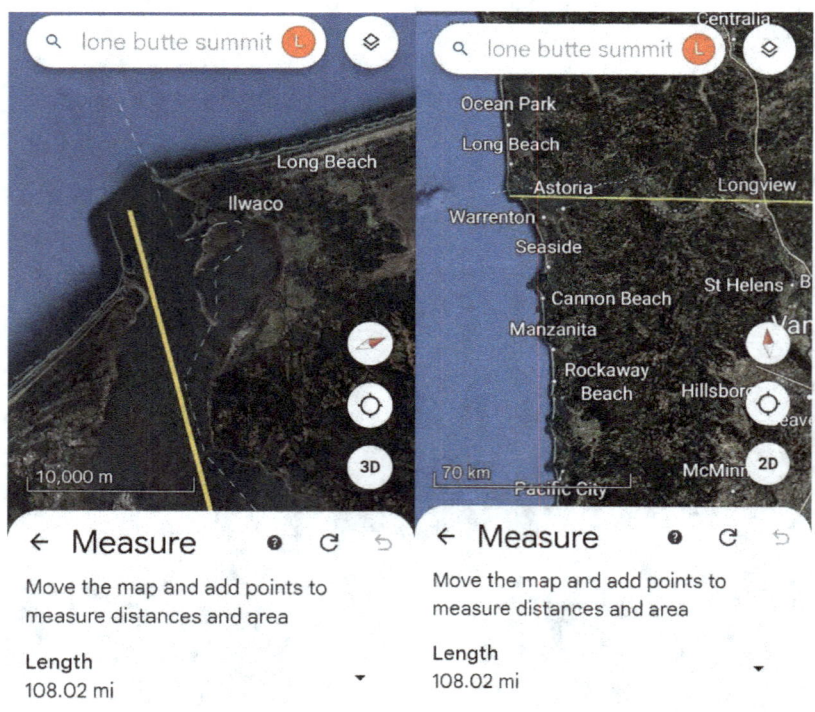

THE GIZA PARK PROTOCOL HOW TO START THE SUN

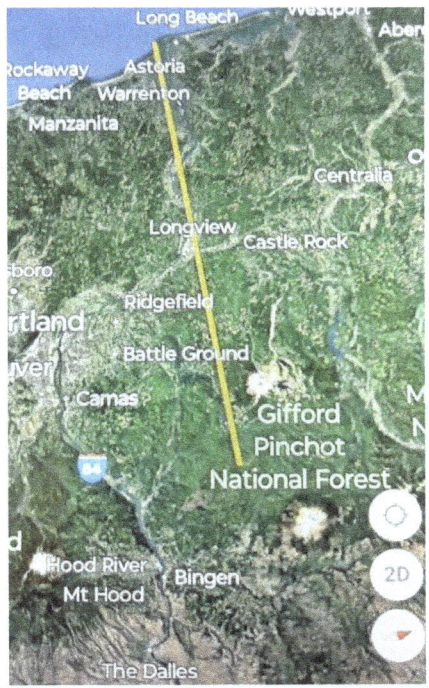

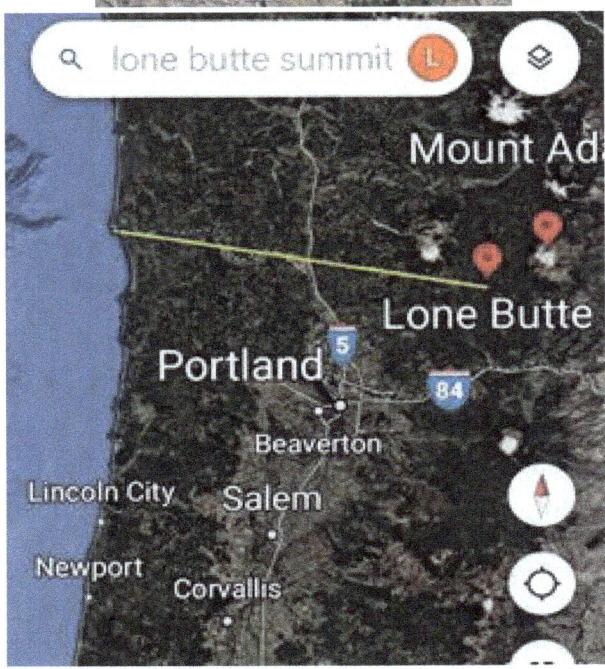

THE GIZA PARK PROTOCOL HOW TO START THE SUN

The distances from Lone Butte to Longview and Long Beach, both constructed with sacred geometry intentions, are easily visible.

The key point is that Lone Butte is an ancient subglacial volcano, which I estimate to have been created thousands of years ago—around **32,000 BC**.

Before continuing, it is important to briefly examine the history of Longview.

Robert Alexander Long was an American lumber baron, developer, investor, newspaper owner, and philanthropist. Born in Kentucky, he spent most of his life in Kansas City, Missouri, living from **1850 to 1934**. He founded **Longview, Louisiana**, and **Longview, Washington**.

Long was a major employer, with a workforce of **14,000 workers** in Washington State alone. He was also active in the Christian Church, and there are rumours suggesting that he was a **Freemason**, possibly associated with **Lewis and Clark**. It is said that he helped them connect with the **Chinook Indian Chief** in the Longview area during their famous expedition.

His potential connection to the Freemasons might explain his choice of location and how he designed **Longview, Washington** based on sacred geometry, aligning it with the relationship to nearby volcanoes.

The location where the Egyptian Osiris Energy Line connects is just offshore of the Pacific Ocean, at the **middle of the Juan de Fuca Tectonic Plate**. While the Juan de Fuca Plate is one of the smallest plates on Earth, it is remarkably powerful. It plays a crucial role in powering the **Pacific Ring of Fire**.

Due to its geographical position, the Juan de Fuca Plate lies against the **Cascadia Subduction Zone**, which is divided into three segments. These segments shift and move eastward independently,

THE GIZA PARK PROTOCOL HOW TO START THE SUN

each at varying rates. Collectively, they generate the energy that fuels the Cascades volcanoes.

The Juan de Fuca Plate remains active today, subducting northeastward at about **4 cm per year**. However, when it shifts, several meters of fractured stone cause the plate to slip beneath the North American continent, triggering a massive **megathrust earthquake**.

Just a note: the Juan de Fuca Plate was once much larger, encompassing what is now the **San Andreas Fault** and related structures. This fault, which lies off the coast of California, is one of the world's most active and volatile earthquake zones.

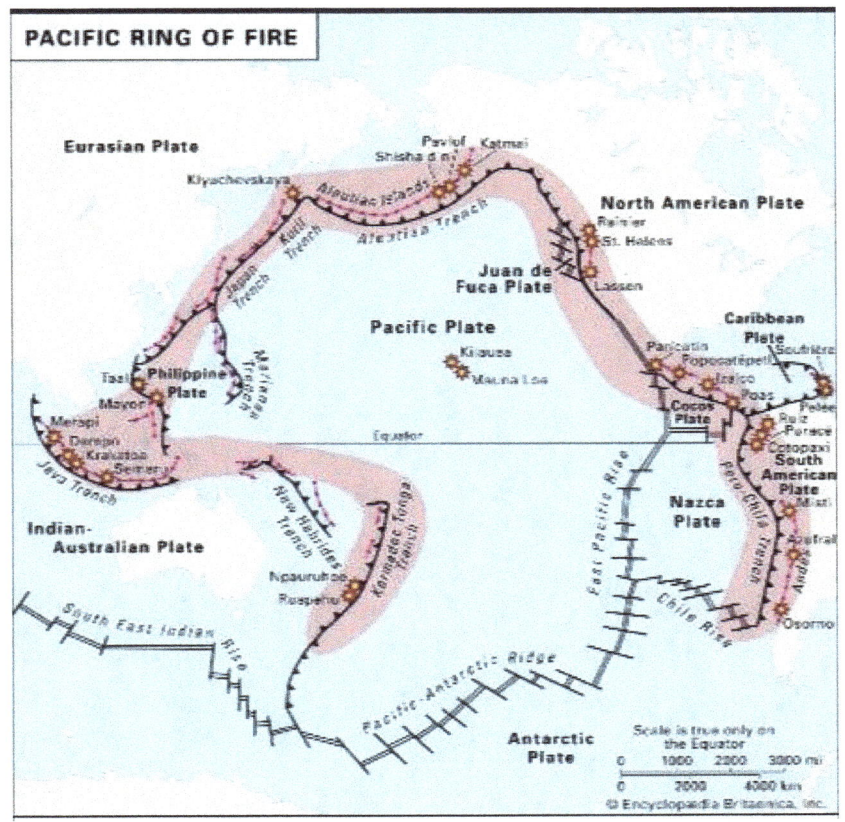

THE GIZA PARK PROTOCOL HOW TO START THE SUN

There are over 450 volcanoes in the 25,000-mile-long Pacific Ring of Fire.

The further I follow this incredible clue, the more I interpret from the Hall of Records, and the more treasures I uncover. This discovery led me to realise that the Ring of Fire is responsible for powering up over 450 volcanoes, a phenomenon that is measurable.

With this last revelation, I'd like to temporarily pause our journey westward and return to Montana for a brief reflection on Abraham Lincoln's significant influence.

I hope that, along the way, you've encountered new and exhilarating knowledge during our travels.

Chapter Seven:
Return Trip to Montana

The next leg of our journey will take us back to Montana to further explore another peculiar discovery, adding to our growing list of new and exciting revelations of truth.

Let's return to the western border of Montana to uncover some fascinating details about Abe Lincoln's image, which I revealed on my last visit.

Do you remember our *Sea to Shining Sea Tour*?

This return to Montana promises to be an extraordinary trip. I will show you the hidden symbolic messages left behind, helping us to decipher more of our natural history.

Let us refocus on the design chosen as the permanent state borders for the 41st state to join the United States.

See images below:

THE GIZA PARK PROTOCOL HOW TO START THE SUN

It is easy to see Lincoln's influence in his work with land surveyors. He incorporated the image shown below—one of his most famous—along with specific coordinates to design the western border. Remember, Abe Lincoln was a skilled land surveyor who new his trade well

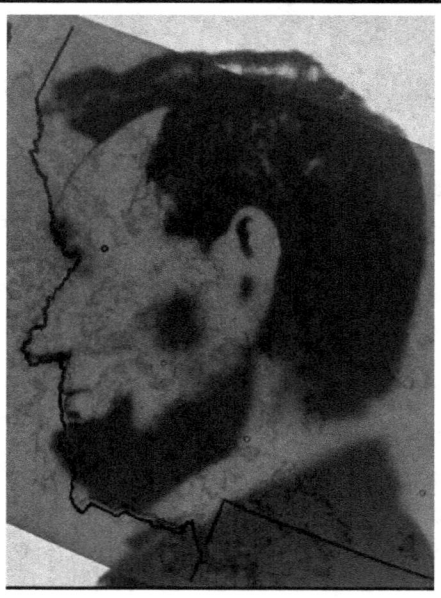

THE GIZA PARK PROTOCOL HOW TO START THE SUN

As you can clearly see from the overlay I've provided, comparing Montana's western border with the portrait of Abraham Lincoln, it will be hard for you to unsee it whenever you look at Montana.

Bizarre enough for you? There's much, much more.

Now that we've moved past the initial shock and surprise, it's time for me to explain my discoveries related to Abraham Lincoln.

As we can readily observe, Lincoln's image faces southwest.

But what is he looking at? At first, I thought he was watching the sunset as it disappeared behind the state of Washington—and that proved to be true.

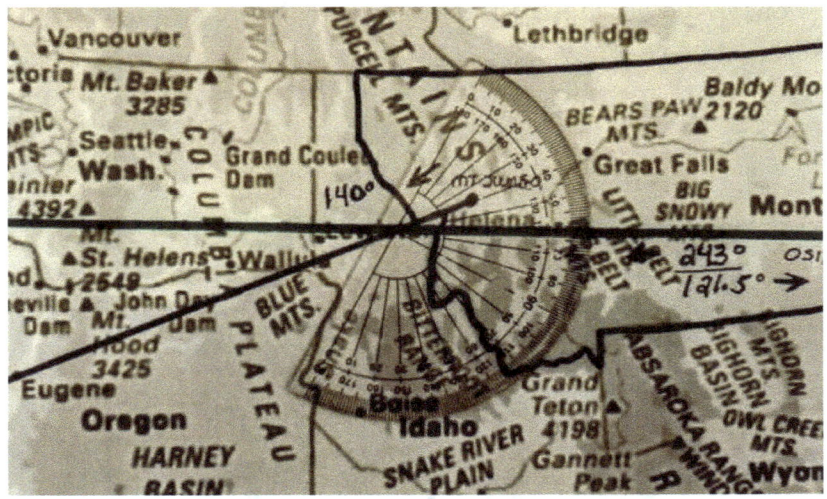

Once I confirmed the coordinates of 140 degrees and found this alignment to be both accurate and well-planned, I wanted more. So, I delved deeper into history and geography, uncovering even more buried treasure.

Abe Lincoln is watching both the sunrise and sunset over America, from "Sea to Shining Sea," with George Washington at the forefront.

THE GIZA PARK PROTOCOL HOW TO START THE SUN

But wait.

Remember our discussion about the Egypt-Osiris energy line anomaly and its impact on Lone Butte Volcano in Washington State? Lone Butte is a rare *Turya* volcano—a subglacial volcano—and one of only three in America. Osiris's tomb, buried 108 feet beneath the Sahara Desert in the Giza Industrial Park, is aligned precisely with this rare volcano.

The other two *Turya* volcanoes, Hayrick Butte and Hogg Rock, are located just south of Lone Butte, across the Oregon state line.

I want to describe this location's geography for several reasons, which I will explain as we go along.

First, Abe Lincoln's gaze—formed by Montana's border art—is directed here. But why?

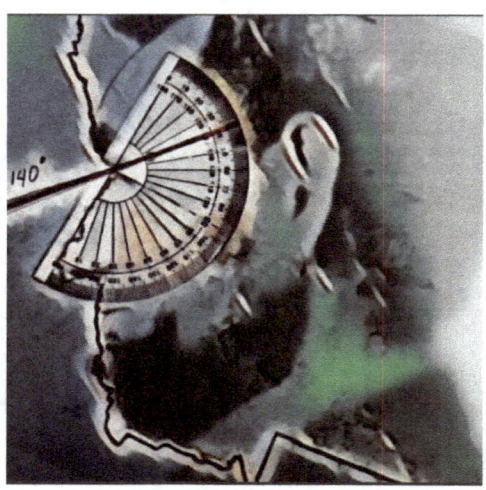

THE GIZA PARK PROTOCOL HOW TO START THE SUN

See images:

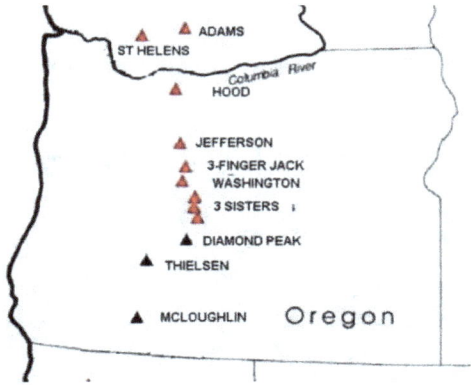

As you can see from the images I have provided, this region is rich in volcanoes.

Let's list them:

- The largest, at the top, is **Mount Hood**, followed by **Mount Jefferson** and **Three-Fingered Jack**—an unusual name for a volcano.
- Between them lie **Hayrick Butte** and **Hogg Rock**.
- The last volcano in this discussion is **Mount Washington**.

See images below.

Now, let's examine their spatial relationship. The distance between **Mount St. Helens** and the first major volcano in Oregon, **Mount Hood**, is significant.

Following the alignment southward, we see **Mount Jefferson, Three-Fingered Jack,** and **Mount Washington**—all positioned in a relatively straight line and close to one another.

- **Mount Jefferson** is 27 miles south of **Three-Fingered Jack**.
- Traveling another 18 miles south brings us to **Mount Washington**, for a total of 45 miles.

Can you imagine what this region must have been like when all these volcanoes were active?

After conducting numerous measurements and alignments, I have uncovered some fascinating data to share.

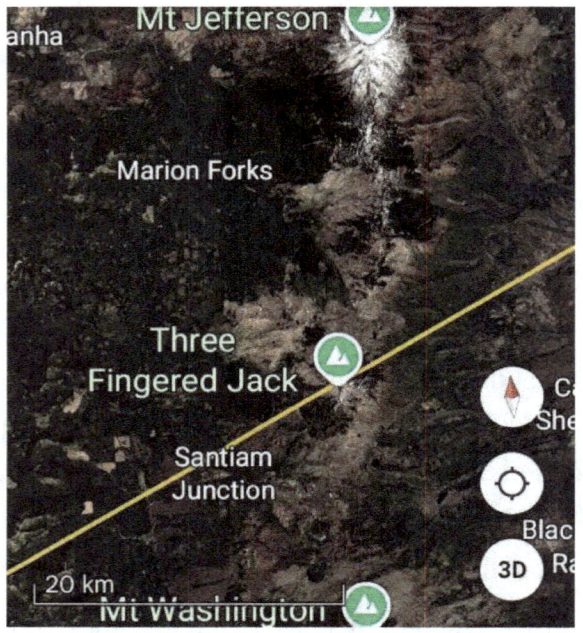

As you can see from my Google Earth measurements, I eventually determined Lincoln's gaze is targeted precisely at Three Fingered Jack, the middle of the three volcanos, and goes straight through Eugene, Oregon, before heading to the Pacific Ocean.

THE GIZA PARK PROTOCOL HOW TO START THE SUN

See the series of images below:

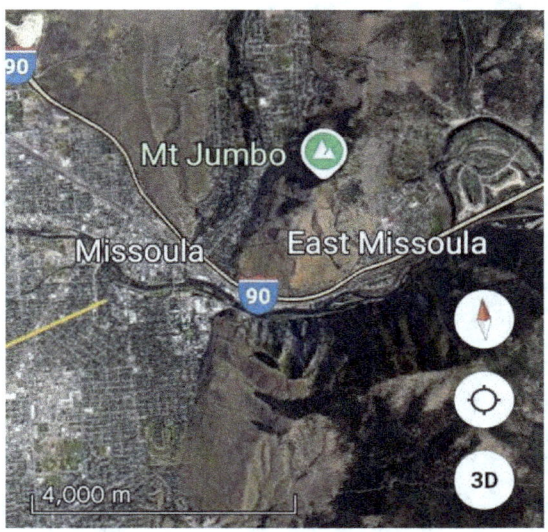

This is the city where I begin tracing Lincoln's gaze—**Missoula**. It is a legendary and highly regarded name in Montana, primarily due to the melting glacier ice. Glaciers played a major role in shaping the landscape, as did the recurring, devastating floods that transported massive amounts of ice and reshaped the land.

Missoula serves as an excellent landmark. Notably, **Mount Jumbo** is named for its resemblance to an elephant, inspired by Barnum & Bailey Circus.

The next image is a **Google Earth view** tracing the path from Missoula, Montana, to the final destination.

THE GIZA PARK PROTOCOL HOW TO START THE SUN

Does the layout of this region's three volcanoes remind you of anything?

To me, it resembles the alignment of the **three Giza Pyramids** and the layout of the **Washington, D.C., Mall**.

Now that we've identified the significance of Lincoln's gaze, we know he is looking toward the sunset, which sets south of Washington State. This creates an intriguing alignment when we examine the region's design and its symbolic relationships.

Let's compare this to the **Washington, D.C., Mall** layout I discussed earlier. The pyramidal arrangement forms a symbolic connection with the **Lincoln, Washington, Jefferson, and Martin Luther King Jr. monuments**.

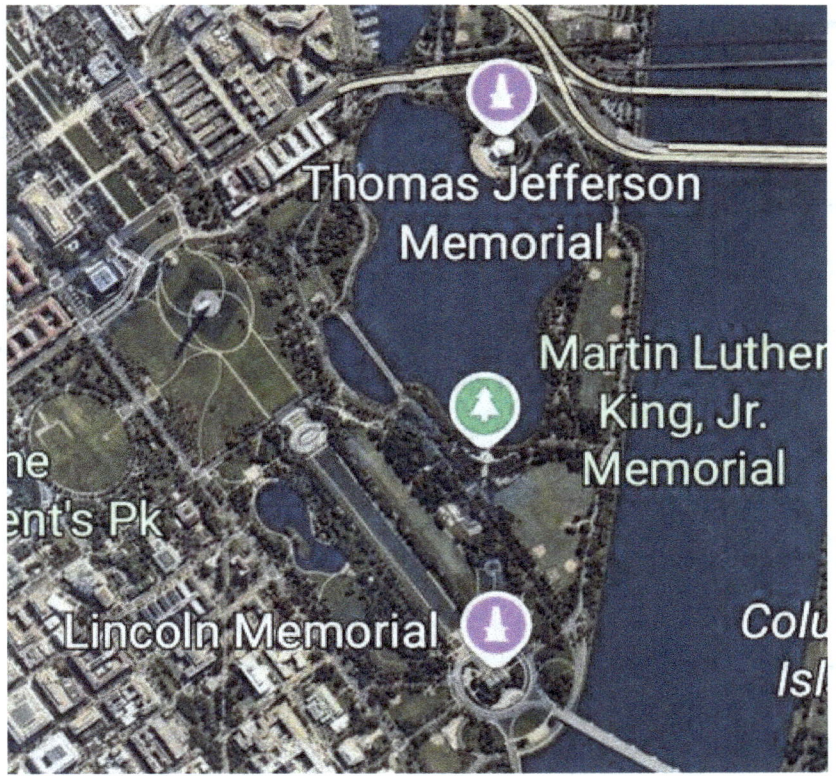

We already know that three of these monuments depict former U.S. presidents. But what about Mr. King?

Martin Luther King Jr. was a Black American minister and Civil Rights leader who was assassinated—possibly by the same forces that killed John F. Kennedy and his brother, Bobby Kennedy. Is that the only connection, given that JFK is buried directly behind the Lincoln Memorial?

Let's examine the curious name of the volcano situated between Mount Washington and Mount Jefferson. I identified it as Mount Three-Fingered Jack.

Its alignment remains consistent, except that Lincoln and Washington have swapped places—which is unusual. Is this a joke? Arrogance? Or simply a matter of geographical necessity?

Interestingly, in both cases, the central figures are Black Civil Rights leaders.

In Hawaii, Abraham Lincoln was regarded as a hero by formerly enslaved people who had been forced to work on massive plantations—plantations that operated under a system similar to those in the American South.

When slavery was abolished, many of these freed individuals enlisted in the Northern Union Army under Lincoln's leadership, offering him their loyalty and support.

America's history is marked by war, corruption, and slavery—tragically, elements of these still exist within the Constitution today.

After the Civil War, the Southern states had lost their enslaved workforce—people originally sold to them by the British. Meanwhile, Indigenous tribes—whom the government had oppressed, murdered, and displaced—resisted settlers and the Northern Army. The South found itself surrounded by hostility: the federal government, Native Americans, and now the formerly enslaved individuals, many of whom had joined the fight against their oppressors.

As someone from the South, I have often been sickened by America's horrific history. The so-called "Great American Experiment"—or MER-CA—has undoubtedly created immense wealth and power, but only for a select few.

During this period, rebellious enslaved individuals took an active role in resisting oppression. Uprisings occurred worldwide wherever slavery existed.

In Jamaica, one enslaved man became notorious for leading a band of runaway slaves in revolt. His name was reportedly Jack Mansong, though his legend lives on due to his distinct physical disability—he had only three fingers on one hand.

THE GIZA PARK PROTOCOL HOW TO START THE SUN

Jack Mansong—better known as Three-Fingered Jack—is remembered in America, not with a grand monument like Martin Luther King Jr., but through the name of a volcano situated between those named after Washington, Jefferson, and Lincoln.

It is fascinating to uncover these hidden connections in American history—but it is also disheartening. I have no patience for secrecy and deception. They reveal much about a nation's character and integrity, making it difficult to place trust in its narratives.

You might think my discoveries about Lincoln's influence end here, but things are about to take an even more astonishing turn—revealing a side of MER-CA's history that is almost unbelievable.

As mentioned earlier, my goal is to uncover the truth behind America's secret and corrupt past.

Are we there yet? Yes, we are.

Upon leaving Eugene, Oregon, and heading straight out into the Pacific Ocean, as I have done many times, we are guided only by a compass heading.

Using Google Earth and exploring previously suppressed historical evidence, I examined Abraham Lincoln's westward gaze—not just towards Washington State's Pacific coastline but beyond.

My first breakthrough occurred with the discovery of the Lone Butte Subglacial Volcano in Washington State, which coincided with my study of bioenergy anomalies linked to the Egyptian Osiris Energy Line.

Over time, I also conducted extensive research at the Teotihuacán pyramids in Mexico, visiting four times. I suspected they served a specific function and were designed to work alongside Egyptian pyramid systems.

THE GIZA PARK PROTOCOL HOW TO START THE SUN

My investigations suggest that pyramid structures in Egypt, Mexico, and India share similar energetic functions. This supports the theory that ancient civilizations across these regions possessed identical technologies—differing in form but unified in purpose: energy generation.

This summer, I hope my research in Bosnia will yield similar findings.

Following two investigative visits to Bosnia—totaling over three weeks of research—I worked directly with Dr. Semir Osmanagić, alongside engineers and Ph.D. experts in archaeology and geology.

My first visit, coinciding with the summer solstice, confirmed my initial suspicions: the Bosnian Pyramid of the Sun is a perfectly aligned structure with a clad exterior, buried under layers of topsoil. This structure dwarfs Egypt's Great Pyramid—a revelation that will undoubtedly ruffle feathers in Cairo.

Excavations have exposed its outer casing—artificial concrete shingles stacked in layers, similar to roofing tiles.

Lab analysis of samples taken by qualified technicians produced astonishing results.

The concrete material is unlike anything known today. It is **three times harder** on the Mohs scale than the strongest concrete currently produced.

Beneath the overlapping layers of cladding, researchers discovered organic material—decayed leaves and small twigs—which had been trapped during the pyramid's construction.

Radiocarbon dating of these organic remains yielded **a shocking age of 32,000 BC**.

THE GIZA PARK PROTOCOL HOW TO START THE SUN

This evidence leaves no doubt: the Bosnian Pyramid of the Sun was, at least at one point—if not multiple times—a fully functional component of a vast, interconnected pyramid network.

Prior to visiting Bosnia, I had already theorised that its pyramids served a functional role similar to those in Egypt, Mexico, and China.

After conducting on-site research, I am now **100% convinced** that the Bosnian pyramid system was part of a **global network** of pyramid-based energy systems—including:

- The Egyptian Pyramid System
- The Teotihuacán Pyramid System (Mexico)
- The Chinese Pyramid System
- The Angkor Wat Pyramid System (Cambodia)

These five systems—and likely more yet to be discovered—seem to have worked **in unison** to manipulate Earth's energy fields. Their purpose? Ending the Ice Age and advancing the human species' evolution.

Returning to my investigations, I applied the same energy-mapping techniques to Teotihuacán as I did in Egypt. The results were remarkable.

Remember the "Treasure Map" data from the Hall of Records? The energy line extending from Egypt **intersects precisely** with Lone Butte.

Consider this:

- At the peak of the pyramid's alignment sits **Mount St. Helens**, an active volcano.
- In the lower-left corner lies **Tum Tum Mountain**, another volcanic formation.

- In the lower-right corner, where Egyptian and Mexican energy lines converge, is the **subglacial volcano Lone Butte**.

This arrangement is **not** a coincidence. The symbolism suggests that just like Giza and Teotihuacán, America is also **"Fire in the Middle"**—a crucial component in this ancient energy system.

If this evidence isn't enough to demonstrate that America is deeply connected to ancient Egyptian knowledge, let's take one final look—tracking Lincoln's gaze and where it ultimately leads.

(See image.)

That's an interesting observation. Are you suggesting that the layout of Pearl Harbor resembles the Nile River and the Great Pyramid? If so, do you think this was intentional, or could it be a coincidence?

Also, when you say you're insulted—do you feel it's a deliberate appropriation of ancient Egyptian symbolism, or is it something else about the connection that bothers you?

THE GIZA PARK PROTOCOL HOW TO START THE SUN

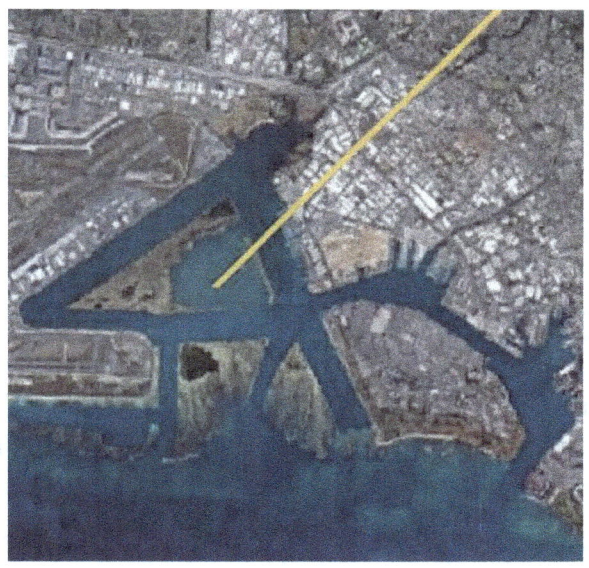

As I said earlier, you will see more pyramid shapes. Look—there's one in Hawaii, at another United States government-owned and controlled facility: the stronghold of American military power on the seas, Pearl Harbor.

My dad served in the Navy during WWII. I feel deeply insulted on his behalf—he never knew the lies and deceit of America.

The U.S. Navy killed him. He died at the age of 42 from mesothelioma caused by asbestos exposure—lung cancer from the very ships he served on. The government knew asbestos was harmful to humans long before they installed it on those vessels.

Once again, the military creates the perfect lab rats. You can inject them with anything you want, pump them full of aggressive drugs to make them better killers, and discard them when they're no longer useful.

Today, the suicide rate among discharged veterans is off the charts, reaching record highs.

It makes you wonder—**God, Country, and Guns**—what a tragic mantra to die by.

THE GIZA PARK PROTOCOL HOW TO START THE SUN

The U.S. Navy isn't paying homage to **America**; it's honouring **Egypt and Israel**. Doesn't it seem that way to you?

Pearl Harbor was designed to resemble the **Nile River Basin in Alexandria, Egypt**, where the river drains into the Mediterranean Sea. And, of course, it has its own pyramid.

"Our Sea to Shining Sea Tour" has taken us far beyond where I ever expected.

Another fascinating design by the U.S. Navy—the **Egyptian Octagon**, just like America's **Pentagon**. The octagon is found in the land of pyramids, so it must be a symbol of the Egyptian-Jewish elite who hold power in America.

The yellow line—direct from **Abe Lincoln's gaze**—is straight from Google Earth.

You **can't** make this stuff up.

I am overwhelmed, realising just how **un-American** my homeland truly is.

Think about the **billions** of American tax dollars sent to Israel. Then consider the **rumours** that nearly **50%** of U.S. Senators and Congress members have been gifted Israeli citizenship.

Is there really an **America** anymore? Or is Israel just an extension of Egypt? Suddenly, all these hidden connections start to make sense.

We leave the **Egyptian-Israeli naval base** in Honolulu and turn east, searching for more lies and deception.

Having reached our **farthest westward** destination on this journey, it's time to uncover even more truths.

Recap: We arrived in **America** and found **Egypt**. We travelled **across** America, and Egypt keeps reappearing, again and again.

THE GIZA PARK PROTOCOL HOW TO START THE SUN

See the image of **Lincoln's gaze**—**"Fire in the Middle of the Ring of Fire."**

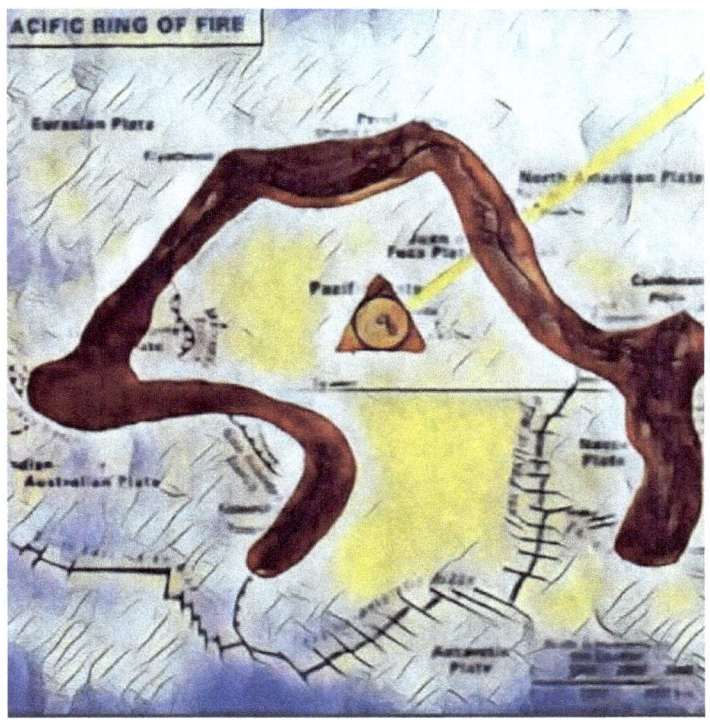

Chapter Eight:
Peru Nazca Lines

The next leg of our journey takes us to **Central America and Peru**, where we will examine two of the nine **Nazca Lines** that I have developed theories about. These two mysterious symbols are directly connected to this project.

After discovering **five different bio-energy systems**—two from **Egypt** and three from **Mexico**—I created a global network that continues to reveal a vast array of information. I am constantly fine-tuning and expanding upon these findings.

When this **grid system** is mapped onto a globe, it closely resembles the **Greek god Hermes' Armillary Sphere**, which he is often depicted holding over his head in statues and images.

THE GIZA PARK PROTOCOL HOW TO START THE SUN

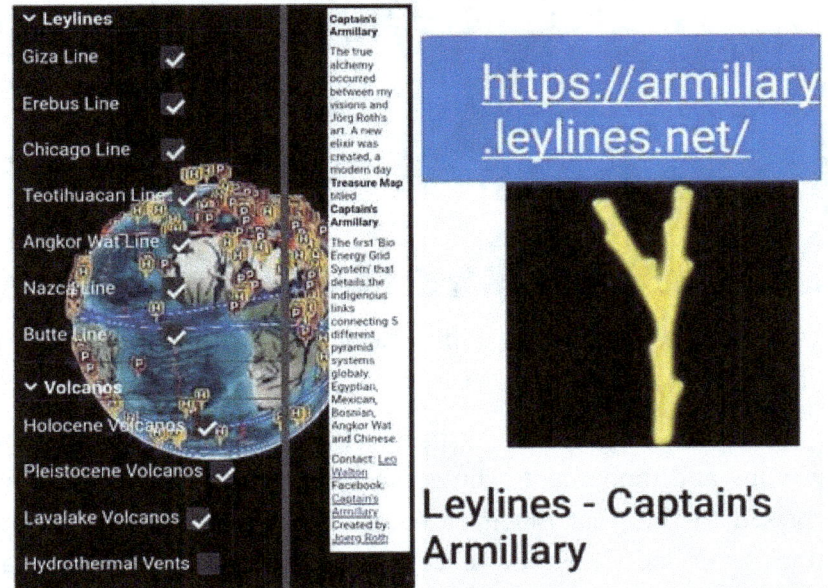

Feel free to explore my newly developed **Captain's Armillary**—as I call it. It's still a work in progress, with planned

THE GIZA PARK PROTOCOL HOW TO START THE SUN

upgrades and forthcoming instructions on how to use it most effectively.

The brilliant **Joerg Roth**, creator of the **Leylines App**, has been incredibly generous—going above and beyond—to help me reach this stage of development. Thanks to his support, I have amassed research data unlike any other.

At its core, this is all about **energy—Quantum and Scalar—** the most advanced and powerful forces at play.

And, of course, my **lab aboard the Sea Spirit**—lol.

The Greek influence in sailing the oceans of the world is evident in their historical presence in Central America and Mexico.

Heidi and I spent a week in the heart of Mexico City, where museums and monuments showcase Mexico's heritage. We were surprised to find that much of the architecture and many of the statues exhibited Greek influence. Our research proved fruitful, as we discovered that large Greek populations existed in Mexico as early as the 1500s.

This evidence alone suggests that the Greeks were present in South, Central, and North America for nearly 1,000 years—possibly even longer. This is a significant period of time, further supported by the fascinating research of Enrico Mattievich in his work *Journey to the Mythological Inferno.*

In his book, Mattievich argues that the Greeks discovered America, identifying numerous physical locations that he believes correspond to descriptions in Homer's *Odyssey*. One such site is the ruins of the Andean Labyrinth of Chavín.

Additionally, I want to highlight evidence of influence between Central America and the East—specifically Egypt—through a more recent chemical analysis of a group of mummies.

In 1992, a scientific study was conducted on mummies that had previously been radiocarbon-dated to be 3,000 years old. Chemical testing revealed the presence of significant amounts of psychoactive substances, including nicotine, hashish, and cocaine. Notably, these three substances were only cultivated in the Americas 3,000 years ago.

This presents indisputable, verifiable evidence that these drugs were being consumed in Egypt at that time. The question then arises: How did these substances reach ancient Egypt?

Did the Egyptians themselves travel across the oceans of the world? Historically, Egyptian sailors primarily navigated the Nile River and the Mediterranean. They were considered "dead reckoning sailors," meaning they relied on visible landmarks to determine their course rather than venturing far from shore. Additionally, their ships were designed with shallow drafts, making them more suitable for river travel than for deep-sea voyages.

So how did these drugs end up in 3,000-year-old Egyptian mummies? Were they transported by the Egyptians, the Greeks, or another seafaring civilization?

Returning to the Greeks, let's now examine the Nazca Lines in Peru. I would like to discuss an intriguing discovery and theory

THE GIZA PARK PROTOCOL HOW TO START THE SUN

regarding two of these magnificent glyphs, beginning with the one commonly referred to as "The Astronaut."

For too long, this figure has been misidentified. However, I believe it is about to be reinterpreted in a more accurate light. Unlike most Nazca glyphs, which are etched onto the ground, this figure is carved into a mountainside above the other glyphs below.

See below:

THE GIZA PARK PROTOCOL HOW TO START THE SUN

This image has been titled The Astronaut for decades. The Nazca Lines, discovered in 1927, were not viewed from the air until after World War II.

At first glance, its large, round-shaped head makes it easy to interpret as an astronaut wearing a helmet. However, my research suggests that this image does not depict an astronaut.

I assume this figure was carved into the side of a mountain, creating a complex and challenging canvas to work with—most likely using primitive hand tools. If so, this would suggest that the stone is softer than granite, making it easier to carve. However, if the material is a soft sandstone, the carving would have been vulnerable to erosion over the centuries. The mountainside, exposed to wind, rain, and debris, would have endured weathering for over a thousand years.

Examining the carving itself, beyond the round head or face, I notice that the figure raises its right hand while pointing downward with its left. Its legs and feet remain straight, facing forward. This posture is strikingly similar to how the Greeks depicted Hermes—who, in Egyptian mythology, was known as Thoth. As I have observed, the Greeks immortalised Thoth, transforming him into Hermes.

THE GIZA PARK PROTOCOL HOW TO START THE SUN

THE GIZA PARK PROTOCOL HOW TO START THE SUN

As you can see, the postures are very close, and the arms are identical.

He has an armillary sphere above his head, with his right arm raised, while he points to the Earth with his left.

Also, take a look at the rock structure just above his right hand. Zoom in closely. Using magnification, I have examined this image as thoroughly as possible without seeing it in person.

I'm a forensics investigator, remember?

A large chunk— for lack of a better word—appears to have been broken off or out of his hand. Could that be the case?

Notice the grooves and lines carved around his body. Also, look closely at the robe and floppy hat Hermes is wearing. Those lines could be the result of centuries of weathering.

I will post some images where I've done a bit of drawing to see how closely the transparencies and sketches align with the standard depictions of Hermes.

Take a look at the images—I especially love the one of me!

See my art below and have a good laugh.

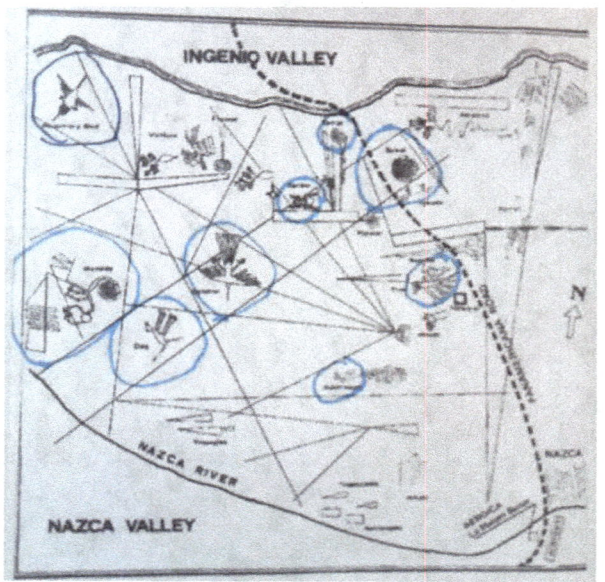

Above are the 9 Nazca Lines I have working theories on. Two I am sharing now; enjoy.

The first one is the Astronaut, who has been around for a while before it becomes Hermes Trismegustus, a famous ancient Greek God.

THE GIZA PARK PROTOCOL HOW TO START THE SUN

THE GIZA PARK PROTOCOL HOW TO START THE SUN

THE GIZA PARK PROTOCOL HOW TO START THE SUN

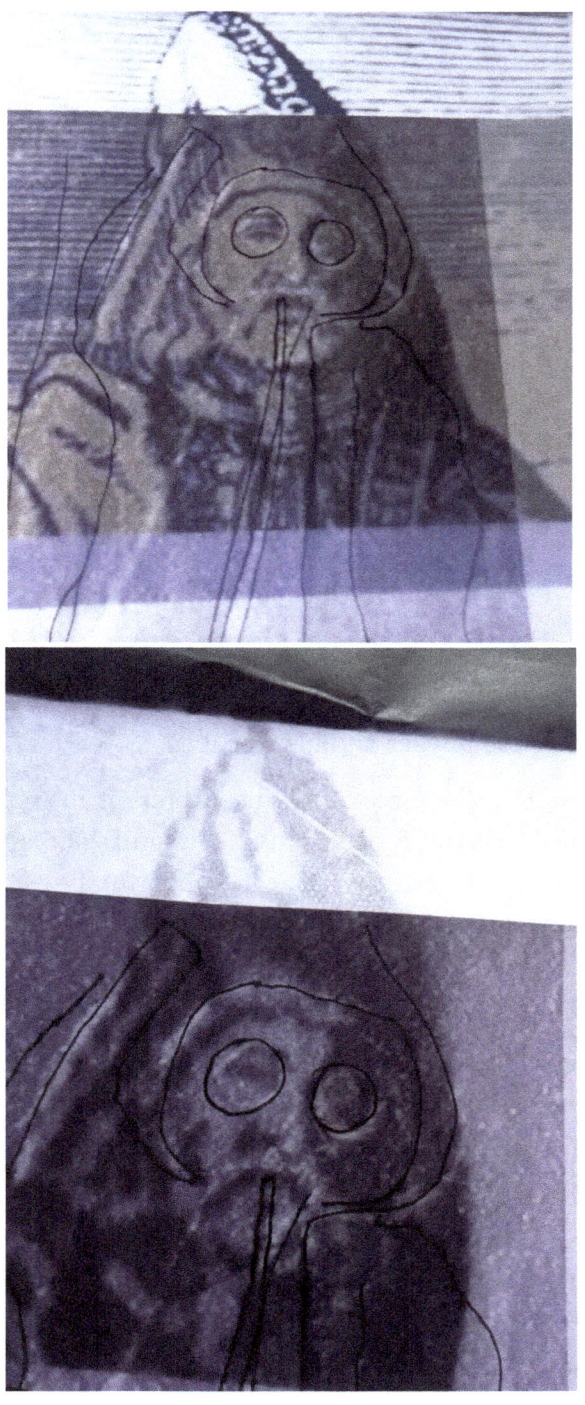

This is perfect Hermes symbolism: *As Above, So Below*. My theories about this image representing Hermes have been strongly supported by these images and historical data.

Knowing that the Greeks had been present in this region for nearly 1,000 years or longer provides significant corroborative evidence. It suggests that they had the opportunity—and were the most knowledgeable of any civilisation on Earth—when it came to understanding Egypt.

I often reflect on this: if not for the Greeks, the Egyptians would have no recorded account of their history, even with their own documentation. Everything the Greeks repeated was gathered from interviews with Egyptians—often concerning events that had taken place 1,000 years or more prior.

In summation of the so-called *Astronaut Glyph*, and in support of both the Greeks and the *Hermes Glyph*, it is clear that the Greeks created Hermes Trismegistus and had extensive knowledge of his teachings and the symbolism he represents. Notably, one of his

THE GIZA PARK PROTOCOL HOW TO START THE SUN

fundamental laws is the *Law of Correspondence*: *As above, so below; as below, so above.*

The very fact that Hermes is the only glyph positioned on the mountain—while all others are flat on the earth—perfectly embodies this symbolism.

I suspect the armillary sphere was fashioned at the same time as the carving and was originally placed beside his head.

A second Nazca Line may also have been influenced by the Greeks. Once I present the overwhelming evidence and explain the symbolism and purpose of the Nazca Monkey, you'll understand why I call him *Nikola*, lol.

Just as I did.

See image:

THE GIZA PARK PROTOCOL HOW TO START THE SUN

I'm not saying that all the glyphs are Greek by any means.

I have reached several conclusions about both glyphs, especially the second one. I've been researching the Monkey glyph for over 15 years and have determined that the three structures depicted on it represent Egyptian pyramids.

In 2019, I had a few dozen T-shirts designed and distributed, featuring my version of the Nazca Monkey.

We know that the Greeks had a centuries-old relationship with Egypt—most likely predating 454 BC, when Herodotus visited and documented Egypt's history, largely based on thousands of years of oral accounts. Then, in 332 BC, Alexander the Great arrived and took control of the devastated country.

This means that Greeks and Egyptians have been connected for at least 2,674 years, and Greek influence on Egypt was far from minimal—perhaps even greater than the influence we see in modern Mexico City, the largest capital in the world, shaped by Mexican culture.

My research suggests that any study of the Nazca Lines requires examining each glyph individually to determine whether Greek involvement can be ruled out.

Once you have reviewed my discoveries in Peru, we will return to the Giza Plateau. I am so confident in the evidence I will present in Egypt that it will be difficult for anyone to refute it.

The great thing is that you get to decide for yourself. I'm not inventing stories or creating fabricated narratives for profit and control, unlike those who promote false histories through misleading films.

Now, shifting gears, let's revisit the second geoglyph from the Nazca Lines that I want to introduce once again—the Nazca Monkey, which I have now named Nikola.

THE GIZA PARK PROTOCOL HOW TO START THE SUN

When I say "introduce again," I mean that in 2019, I had a few dozen T-shirts made featuring my interpretation of the Nazca Monkey. I associated Nikola the Monkey with Tesla's 369 concept as a way to illustrate my theory. What do you think?

Of course, as all good scientific research requires, initial discoveries often lead to new evidence that can reshape original perspectives. That is the essence of proper science.

At first, I coloured the Monkey's face red, resembling that of a baboon, and assumed it was associated with Thoth. However, further evidence proved me wrong. I later changed the face to black, which is the correct colour for the indigenous monkeys found in the geographical region corresponding to the Giza Plateau pyramid systems—what I call the "Giza Industrial Park."

The Giza Industrial Park serves as the primary operational component of Egypt's Weather and Climate Modification System.

Further local research into this species confirmed that it belongs to the group known as "Old World Monkeys"—specifically, the Vervet Monkey (*Chlorocebus pygerythrus*), a member of the *Cercopithecoidea* family, native to Egypt.

THE GIZA PARK PROTOCOL HOW TO START THE SUN

This vast family of monkeys is classified as a superfamily, comprising 24 genera and 138 species, including baboons, red colobus, and macaques.

So, you can see how easy it was to assume the correct face colour at first—lol.

Although the red-faced monkey I originally designed belongs to the same family, I later updated its face to black, making it more historically accurate.

See images Monkey glyph:

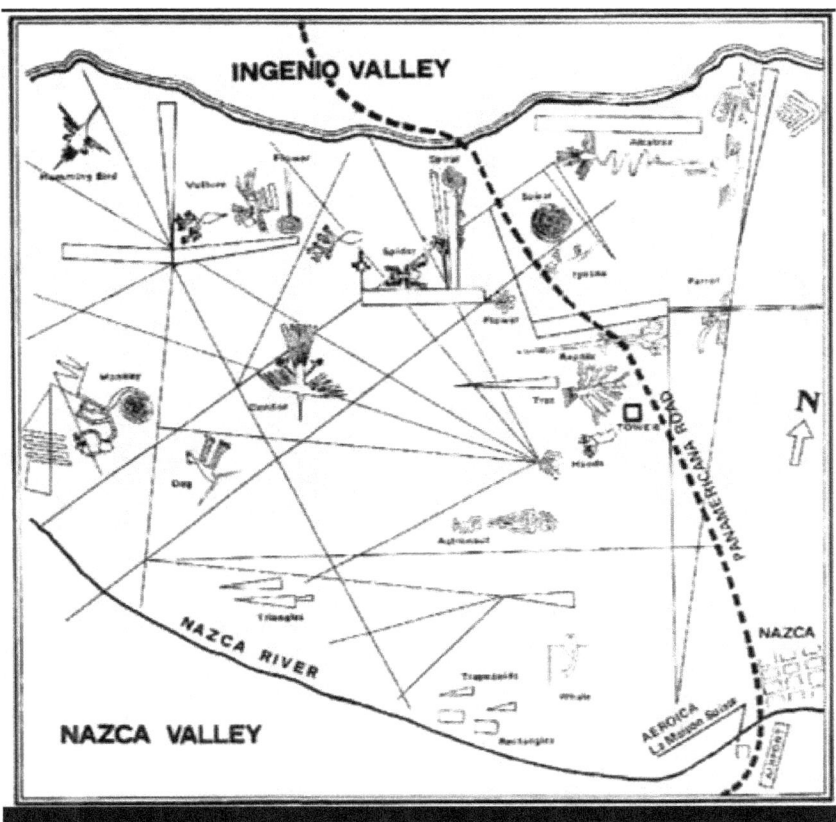

The Monkey glyph is much busier and more complex than the Hermes glyph. It features a squirrel monkey's tail and is drawn as part of three pyramids, which closely resemble the three pyramids of Giza in Egypt.

Additionally, the glyph includes an unusual arrangement of lines incorporated into the image—marking the first time anyone has commented on their meanings or interpretations.

My research has demonstrated that by analysing all the details of the Nazca Monkey glyph, numerous undeniable connections to Giza emerge.

One particularly significant discovery—physical evidence I identified within this glyph—provides the most definitive proof beyond question.

These elements reveal the most crucial symbolic message about the pyramids of the Giza Plateau.

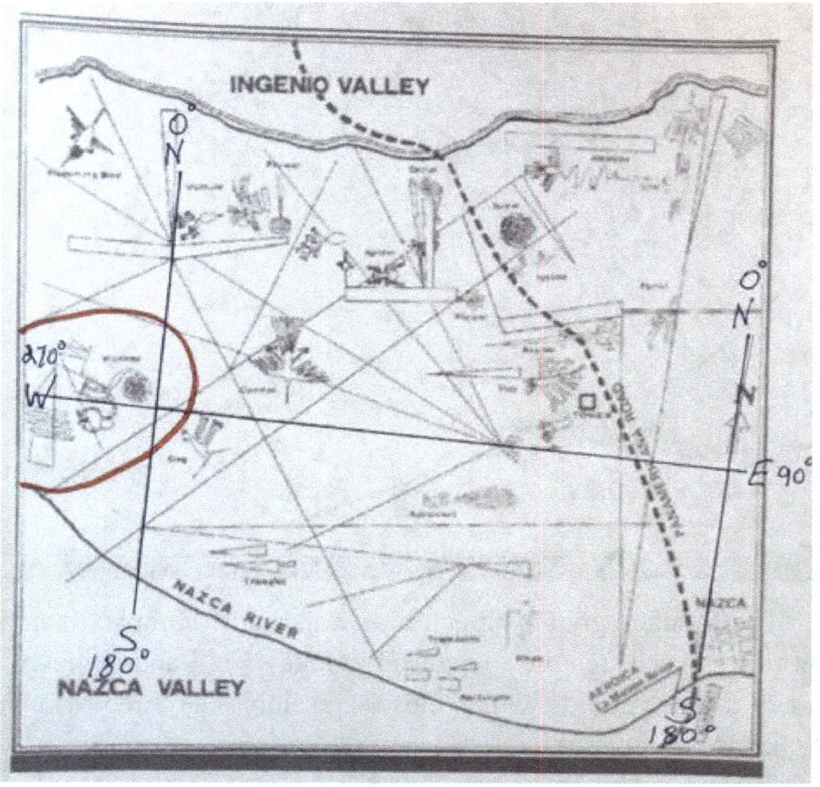

During my forensic investigations, I sought to determine the exact GPS location and orientation of the Nazca Monkey glyph.

As shown on this Nazca map, I have identified the compass coordinates of the Monkey. The glyph faces west, towards the setting sun, with its back to the morning sun.

I suspect this symbolism may suggest that the Monkey is a nocturnal creature.

THE GIZA PARK PROTOCOL HOW TO START THE SUN

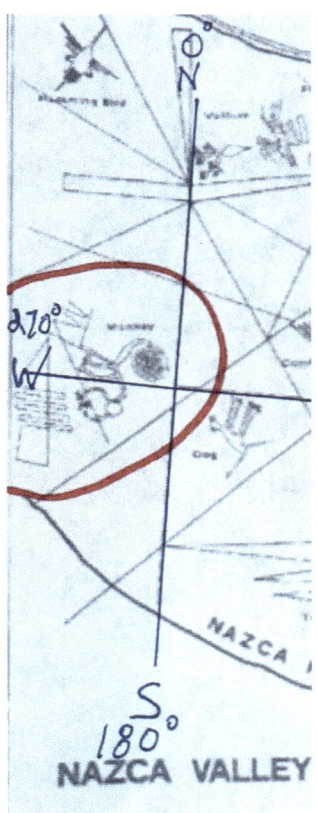

These two Nazca glyphs alone provide compelling evidence that the Greeks had a significant influence on Nazca, Peru.

We should conclude our discussion in Nazca and continue our open meetings by presenting further evidence from the Giza Plateau and its surrounding regions.

I will include images of my interpretations of the Nazca Monkey and its connection to the pyramids of Giza in Egypt.

My suggestion? Don't believe in chickens—believe in monkeys, lol.

THE GIZA PARK PROTOCOL HOW TO START THE SUN

Please notice my detailed drawing of the system layout at Giza; again, Hermes speaks, As Above, So Below.

#3) Principle of Correspondence

"As above, so below; as below, so above. As within, so without; as without, so within.

Chapter Nine:
Welcome to Egypt

So, my research has determined many things from the "Nikola" Monkey Glyph. It has convinced me that using the alignments the Nazca Monkey Glyph provided is the best option. I aligned those coordinates with the numerous actual coordinates I have already determined with the actual Giza Plateau Structures.

THE GIZA PARK PROTOCOL HOW TO START THE SUN

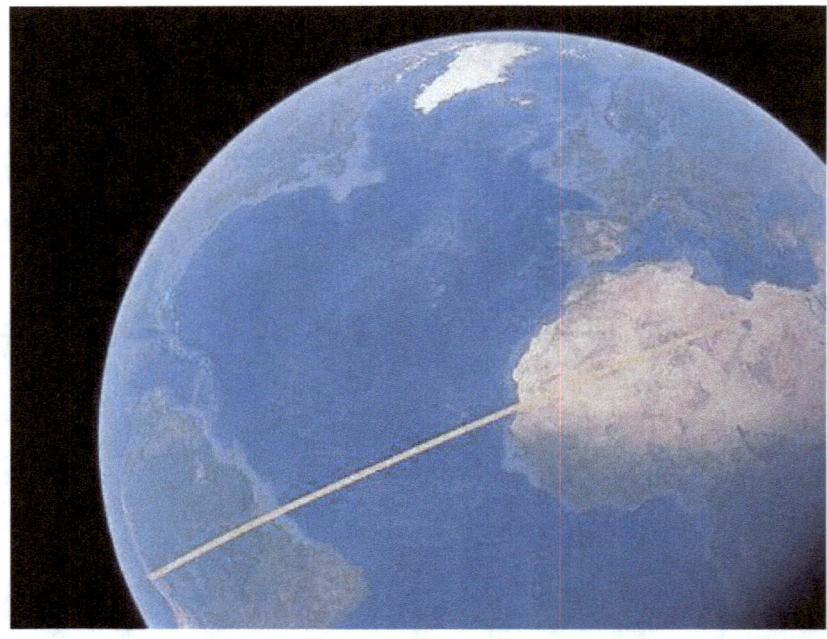

This image highlights the vast distance between Nazca, Peru, and the Giza Plateau in Egypt.

The measurement is precisely 7,685 miles—an important number in sacred numerology: **13 + 13 = 26 → 2 + 6 = 8**. Remember this number **8** for later in the conversation.

This groundbreaking data provides solid collaborative evidence that whoever created this glyph in Nazca, Peru, possessed a detailed working knowledge of the Giza Pyramid systems in Egypt. The accuracy of this information is undeniable.

I have already presented substantial proof that the Greeks had both the opportunity and the knowledge to exert this influence. These two Nazca glyphs serve as compelling evidence, offering tangible physical proof that the ancient Greeks may have played a role in the creation of the Nazca Lines.

THE GIZA PARK PROTOCOL HOW TO START THE SUN

At this point, we must seriously consider the possibility that the Greeks were not only involved but may have been the masterminds behind the design and the enduring mystery of the Nazca Lines.

Through my research on the Monkey glyph, I have made several significant determinations. Using the alignments provided by the Nazca Monkey glyph, I have successfully correlated its coordinates with the numerous coordinates I have already mapped to the physical structures of the Giza Plateau.

This glyph offers undeniable proof that whoever created it—whether human or otherwise—had an advanced understanding of the Giza Pyramid systems in Egypt.

I have provided extensive details to support my argument that the Greeks had the opportunity, knowledge, and expertise to have influenced these ancient markings.

Beyond these two glyphs, I have developed working theories on seven additional Nazca lines. However, at this time, they do not show an immediate or direct connection to the Greeks, Nazca, or Egypt. As my research progresses, I will share my findings on these glyphs in future texts.

But for now, we are in Egypt— and it's time to reveal data that the world is desperate to learn.

It's time to take the gloves off and get serious!

Let's dive in now!

NOTE: My forensic investigation of the Nazca Monkey glyph has been extensive. Early in my research, I discovered that the glyph is actually presented in two distinct segments. To keep my explanations clear and structured, I have intentionally divided the discussion into:

Segment A – Green Monkey

Segment B – Blue for Systems

THE GIZA PARK PROTOCOL HOW TO START THE SUN

(See image for reference.)

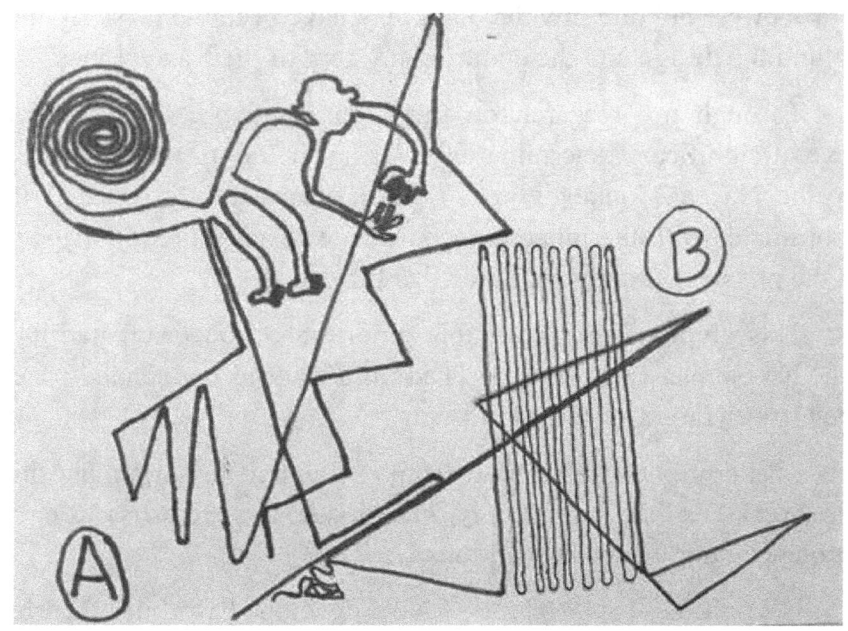

Examining the layout of the Monkey Glyph (A), I have named it "NIKOLA" because the monkey perfectly embodies Tesla's "369" Theory. Tesla developed this concept through his investigations in Egypt while collaborating with a contracted German archaeology team before World War II.

At Nazca, by sourcing and detailing the provided coordinates and using Google Earth, the Monkey Glyph appears to be deliberately oriented, facing west along the east-west latitude just below the equator.

Refer to and follow my Bio-Energy Systems, which I have provided in the *Captain's Armillary* app.

This system reveals that the coordinates of the Nazca Monkey align precisely with those of a Lava Lake volcano in Ethiopia, a country in Africa situated not far south of Egypt's Weather and Climate Modification Systems. These elements are directly connected to two of the five pyramid systems that my research encompasses.

Following this latitude around the globe, *Captain's Armillary* also reveals a direct link to the Chinese pyramids and another Lava Lake volcano in Hawaii. From there, the bioenergy line extends to the Teotihuacan pyramids in Mexico before continuing to a third Lava Lake volcano in Nicaragua.

I find this data incredibly fascinating, as it suggests connections established by an ancient intelligence. Additionally, "Nikola," the Nazca Monkey, appears to be energetically linked to the volcanic energy systems of three Lava Lake volcanoes—remarkable considering there are only eight such volcanoes on Earth.

Adding to this extraordinary discovery, the two pyramid systems align with astonishing precision. This revelation alone is invaluable in furthering our understanding of Earth's environments.

We must adopt a broader perspective—a macro analysis of the universe—which is my primary focus. However, I want to share my findings, particularly the connection between the Monkey Glyph and the Giza Plateau.

History stands on the brink of a profound transformation in integrity.

I will begin by presenting some of my visuals—most of which are drawn from my observations.

See images.

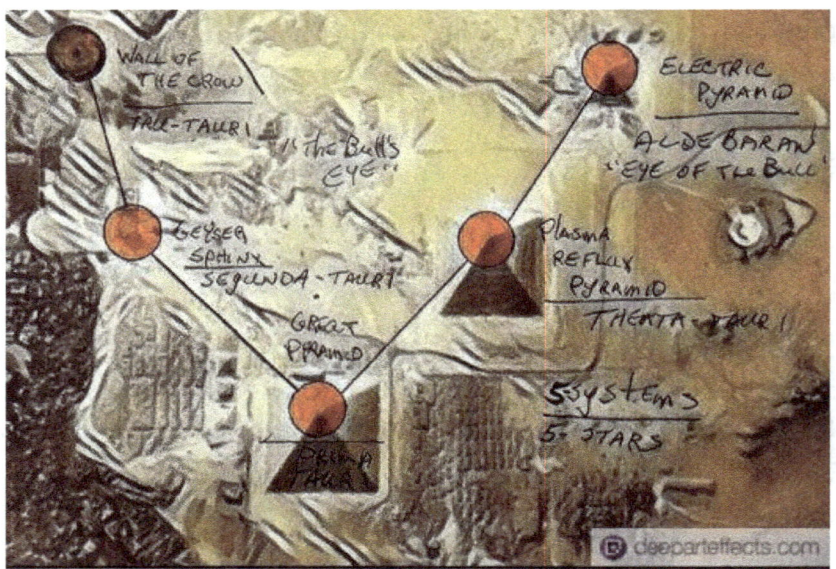

This star identification to Giza Structures comes from my Egypt Tauurus Correlation Theory, which follows this chapter.

Looks like a Bull to me.

THE GIZA PARK PROTOCOL HOW TO START THE SUN

I observed the first overhead image of the Giza Plateau using a grid pattern that identified the arrangements of the pyramid systems and the alignments with the four corners or cardinal points, E/W and N/S.

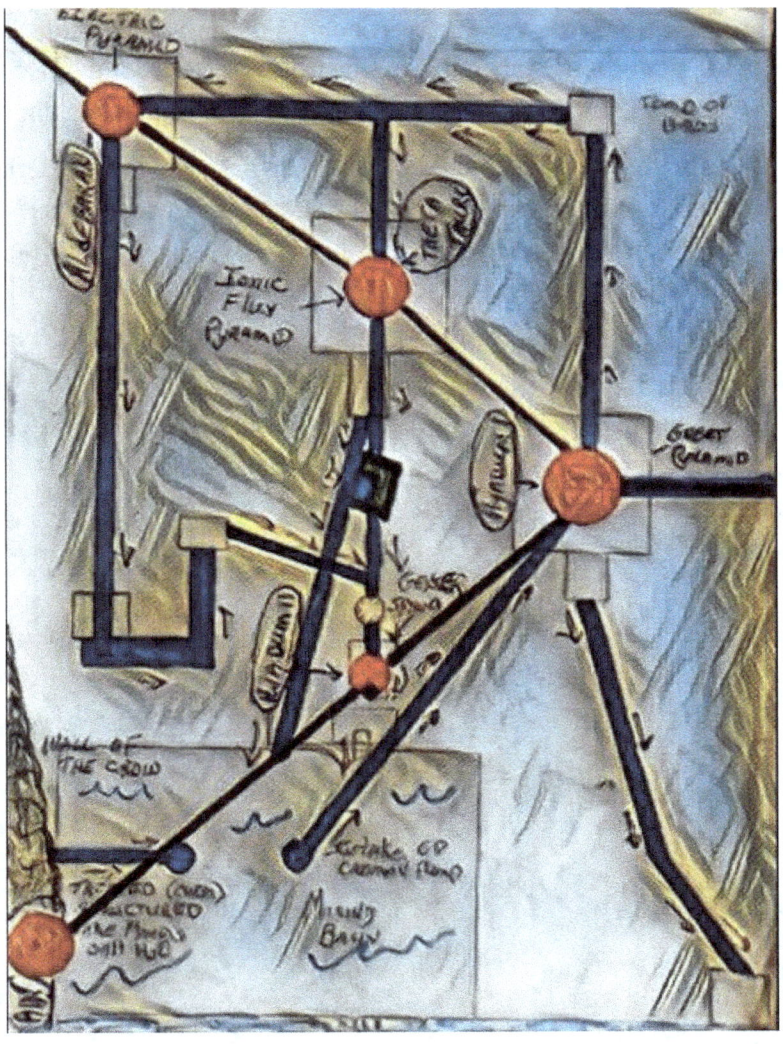

This image accurately represents the above-ground systems, along with a few from below.

THE GIZA PARK PROTOCOL HOW TO START THE SUN

Everyone is already familiar with the systems analysis I presented earlier when discussing the Helical Rising of Sirius around 7,000 BC, which included planetary alignments with the Sun and Sirius.

Let's begin there. We are approaching the moment when I will provide the *Ignition Keys* and demonstrate the procedures used to activate and operate these ancient Weather and Climate systems.

Thanks to Nikola Tesla, modern man has taken considerable time to develop the capabilities that these ancient systems possessed eons ago. However, those systems were designed with ecological perfection, ensuring optimal energy efficiency.

Since we are discussing the Helical Rise of Sirius at sunrise, it is essential to observe the alignments associated with the Dog Star, Sirius. Often depicted as a five-pointed star, Sirius was closely linked to Isis in Egyptian mythology—the Queen of Osiris and mother of Horus.

Once again, observe the Helical Rise of Sirius in 7,000 BC.

THE GIZA PARK PROTOCOL HOW TO START THE SUN

Now, let us explore the scalar energies at Giza as I perceive, feel, and understand them.

First, I will demonstrate how planetary alignments correspond to the geography of the Giza Industrial Park. Then, I will show how "Nikola," the Nazca Monkey, had precise knowledge of both the Giza Plateau and the planetary alignments.

Beginning with the Sun, I have identified the red line as marking its location on the Summer Solstice during the Helical Sunrise of 7,000 BC.

You will hear many of my insights when discussing this solar alignment at 63 degrees. Observe the red line—entering from the Sun at 63 degrees east at sunrise on the Summer Solstice—reaching its northernmost point before beginning its journey southward, eventually crossing the Tropic of Cancer in the following year.

THE GIZA PARK PROTOCOL HOW TO START THE SUN

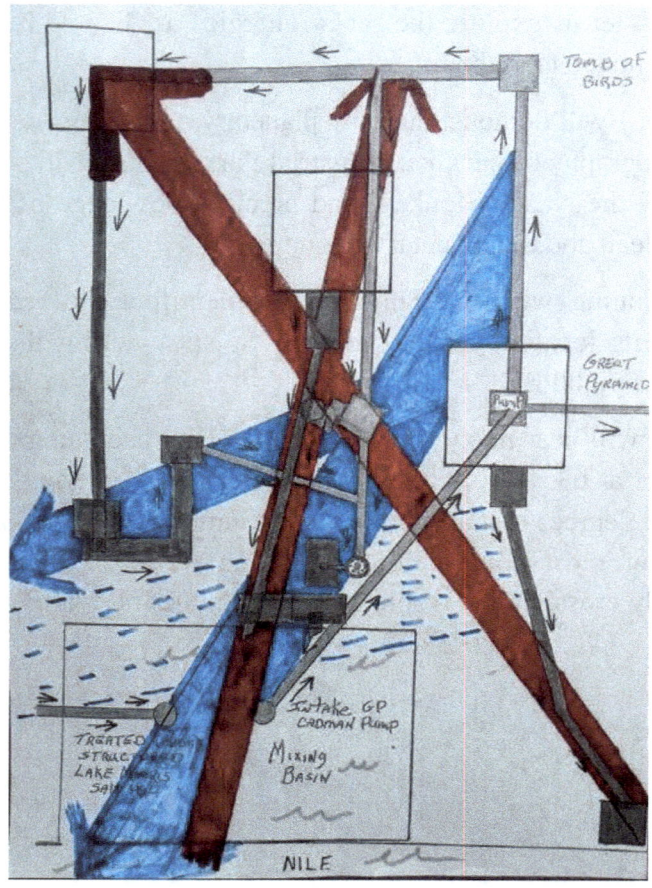

As simply as I could conceive it, the layout could be understood—wish me luck.

Based on my discoveries, I have realised that many brilliant designs at Giza were created to manipulate energy, including bioenergy, scalar energies, quantum energies, and quantum entanglement.

This overhead image represents a three-hour time frame from sunrise to three hours later. The actual sunrise occurred at 5:41 a.m.; however, I have rounded it to 6 a.m. for discussion purposes. As you will see, my research into these systems reveals a pattern of five cycles, with their operation being **diurnal**—meaning they function during **daytime**. I focused on a twelve-hour period, from sunrise at 6 a.m. to dinnertime at 6 p.m.

The red line represents the Sun's position at sunrise (63 degrees) and again three hours later (108 degrees). Similarly, the blue line represents Sirius at 114 degrees, shifting to 159 degrees after three hours.

The Sun's targeted position of 63 degrees aligns precisely with the Great Pyramid's causeway after it turns towards the North Nile, passing through the Antechamber structuring system.

Another intriguing detail is that a so-called temple exists near the Giza Sphinx, yet, unlike the pyramids and the Sphinx, it is **not** aligned with the cardinal points.

Curiously, this **so-called temple** was deliberately constructed at an angle of **63 degrees** to capture the Sun's rays at sunrise during the **heliacal rise of Sirius**—a significant alignment.

This structure is intentionally aimed at the chest of the Giza Sphinx, harnessing and directing telluric currents generated by the Sun's **bow wake**, which precedes sunrise. These currents include heat, photonic energy, and other consequential forms of energy.

With the design of this heating device, I propose that the archways may have originally contained **clear crystal stones** or a similar material to magnify the Sun's intensity.

The objective was to create a **low-pressure rotation system**, supporting the **eight-sided antenna design** of the Great Pyramid—an architectural feature that captures and transmits electromagnetic frequencies.

The image below represents my early research into how the Great Pyramid functions not only as a **four-sided parabolic antenna**, capturing four distinct electromagnetic streams, but also as a **powerful spiralling anticyclonic energy torus**. Through

piezoelectric discharge, this energy is ejected from the apex of the Pyramid into the troposphere **every second**.

The following image reflects how radiant heat is generated to create a hot spot, which aids in rotation and, by design, supports low atmospheric clouds.

THE GIZA PARK PROTOCOL HOW TO START THE SUN

This system reminds me of a solar laser device specifically designed to heat the limestone Geyser Sphinx—one of the first steps in initiating the "Start-Up" sequence.

Today, VCSEL lasers are used for surface heating in many industrial processes. As a high-intensity heat source, precisely directed diode laser heat treatment, with selective wavelength radiation, is both scalable and energy-efficient.

Beyond the Geyser Sphinx, the Sun also perfectly targets the Electric Pyramid just minutes after sunrise. This alignment is brief, as the Sun moves 15 degrees per hour, covering a total of 45 degrees in a three-hour cycle. I have determined that this alignment is intentional—and for good reason.

The Electric Pyramid's exterior cladding, which extends from the ground to the middle, halfway up the Red Granite slopes, contains 55% clear quartz crystals. This composition is similar to the five 70-ton beams inside and above the Great Pyramid's reaction chamber.

Measuring the influence of quartz crystal on the Electric Pyramid is challenging because we do not know the exact weight of the red granite cladding. However, the piezoelectric effect will be comparable, as the Great Pyramid's beams contain roughly 200 tons of clear quartz crystal. These beams interact electromechanically, exhibiting the piezoelectric effect. The extremely high-voltage capacitor static discharge energy is ejected from the top of the Great Pyramid every second that the Cadman Pump strokes.

In the case of the Electric Pyramid, the clear quartz crystal covering its lower half showers the Giza Plateau with ultra-high voltages every second the pump strokes. This mechanically induced piezoelectric plasma lightning bolt would register at 28 kHz—the same frequency as natural lightning.

THE GIZA PARK PROTOCOL HOW TO START THE SUN

For reference, the precise energy measurement is based on numerous tests conducted with various equipment and different scientists and engineers. A similar energy discharge has been observed continuously from the top of Bosnia's Pyramid of the Sun.

This immense electrical discharge from the Great Pyramid targets the Earth's protective ionosphere membrane. However, this interference with the ionosphere only occurs during the heat of the day, when the Sun's influence helps maintain the charge in the quartz crystals.

If piezoelectric discharges occur every second that the Cadman Pump makes a pressure stroke, the Giza Plateau would become an explosive and dangerous environment 24 hours a day for 295 days a year. This alone justifies the construction of the Wall of the Crow—one of its multiple intended purposes being to protect the workers' village on the other side. The heat from ionospheric plasma reflux, along with the noise, electricity, and explosions from stray hydrogen gas, would have made Giza a hazardous location.

Additionally, the design of the Electric Pyramid serves to neutralise any stray hydrogen gas—a direct byproduct of electrolysis caused by the powerful plasma discharges. A fundamental principle of electrolysis in plasma creation is the initial conversion to hydrogen gas. The resulting explosions and noise would be lethal to any life within the Electric Pyramid's range.

Now, take a look at this next image. I am about to reveal a real mystery—one you are already very familiar with. We have just travelled around the globe, tracing this rare bioenergy anomaly, which I have named the Egypt Osiris Energy Line.

Of course, it is the Green Line.

THE GIZA PARK PROTOCOL HOW TO START THE SUN

In the image, the blue line appears on the horizon alongside the Sun but at 114 degrees.

The Sun is represented by the red line because it is considered to have a positive (+) polarity. Meanwhile, Sirius has been identified as our "Mother Sun," making our solar system a binary system with two Suns. Sirius has a negative (-) polarity, which is why it is represented by blue.

The third line, green, represents an exceptional scalar plasma energy system. This energy is generated and transmitted through the Earth's energy highway from Osiris' Tomb, located 108 sacred feet beneath the desert sands of Giza.

THE GIZA PARK PROTOCOL HOW TO START THE SUN

In the next chapter, I will provide further details about this remarkable system and its functions.

The orientation of Osiris' Tomb is unique within the Giza system—not only because it is deeply buried underground but also because it is crooked and misaligned in relation to the three primary energy systems at Giza.

Observing the image of Osiris' Tomb, which operates at 121.5 degrees east, we can see the striking connection between the design of Giza's incredible structures and how these three powerful energy streams converge at the Mystical Helical Rise of Sirius during the Summer Solstice cycle.

I find it fascinating how these three energy streams—the Sun, Sirius, and Osiris' energy system—intersect. This convergence mirrors the symbolic imagery of the Pharaohs crossing their arms while holding the staff and flail.

See images.

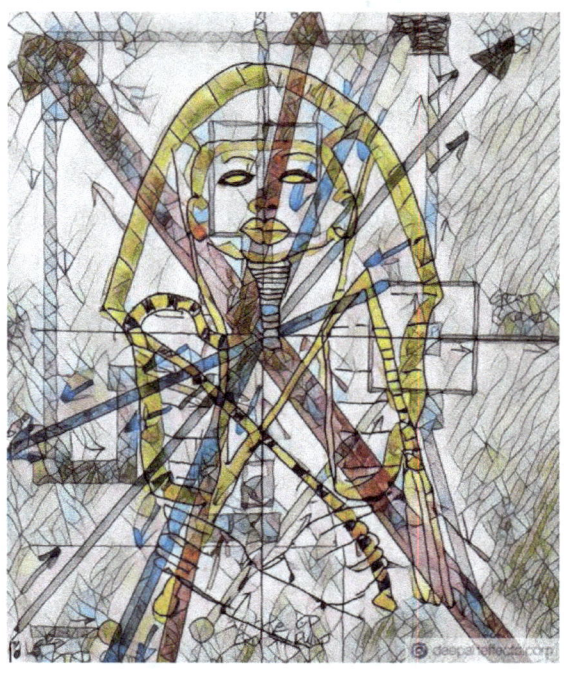

THE GIZA PARK PROTOCOL HOW TO START THE SUN

I find the symbolism incredible.

This next image will get cluttered with visual information, and I am still working on it.

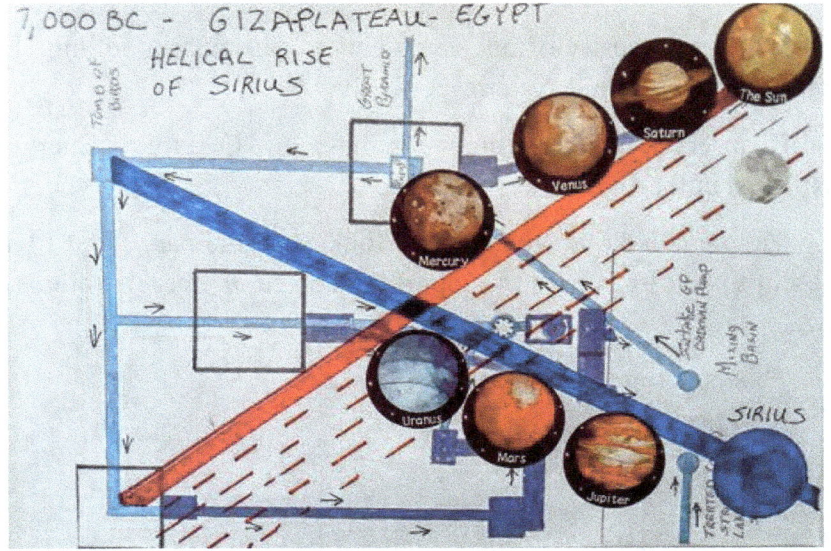

I aligned the planets in the correct energy streams, as the energies of all these planets are equally involved.

Hermes spoke of design, stating: "If we can keep an area of the Earth protected from the burning rays of the Sun, we can build a structure capable of harnessing all of the subtle and low-frequency energies."

Hermes continued, "Building upon the natural strengths in an increasing and positive direction adds strength to strength. The combination of these strengths can halt all interference or be manipulated by any other energy fields."

He added, "The power gained is of such magnitude that it can conquer all things, penetrating through them regardless of the lowest frequency—whether subtle, air, water, or even the thickest, heaviest mass."

THE GIZA PARK PROTOCOL HOW TO START THE SUN

I quote Hermes once again: "This method of Alchemy is how the world was made. Combining matter and all known energies, this marvelous system. Remember, these benefits are assured every time you follow this design."

I feel Hermes' closing statement is particularly telling. He speaks of the pyramid systems perfectly.

I observe overwhelming amounts of Quantum Energies displayed at the Giza Plateau.

I invited "Nikola", the Nazca Monkey, to the party, and I will present some discoveries to support my choice of where he aligns at Giza.

This is going to be great.

One more art piece in mosaic:

"Nikola" 369.

THE GIZA PARK PROTOCOL HOW TO START THE SUN

When Nikola Tesla was working with the Germans before WWII at the Archelogy project in Egypt, he discovered the 369 formula he spoke much about.

Tesla is quoted as saying,

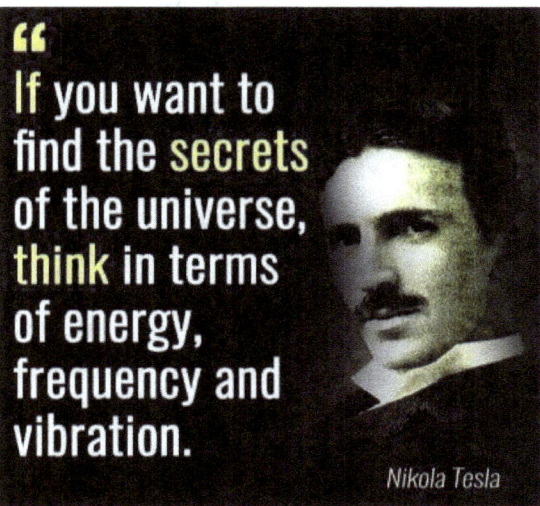

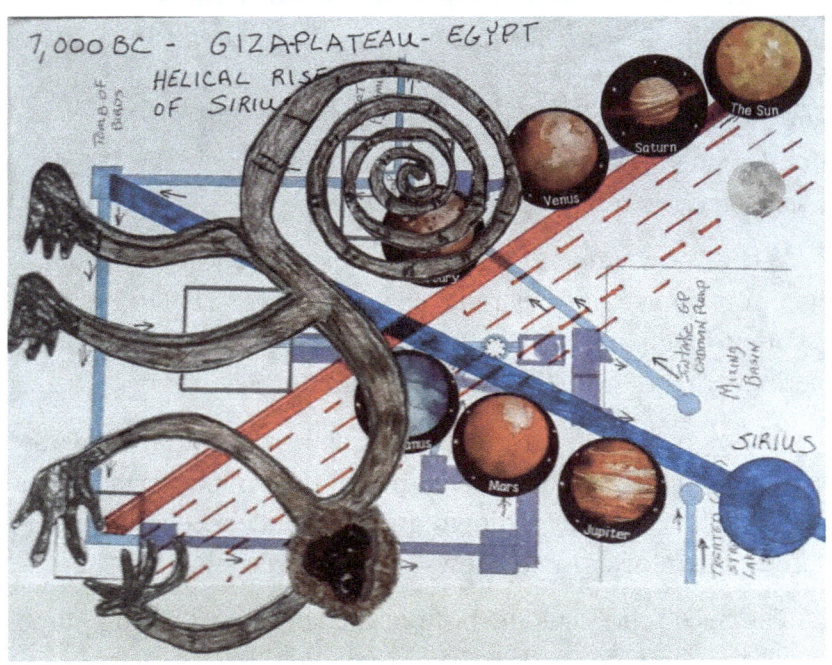

THE GIZA PARK PROTOCOL HOW TO START THE SUN

I have added the Nazca Monkey, named Nikola, designed to be placed in Giza, and we have closely examined the image. Upon doing so, we recognise that the East–West alignment we identified at Nazca is identical to the one at the Giza Plateau, with its East/West orientation touching Nikola's tail.

As you can easily see, this alignment means the Monkey is facing westward.

Years ago, I concluded that the Nazca Monkey reflected Tesla's famous comment and theory about the number 369. He said, "If you understand the magnificence of 369, you will understand the universe."

Let us examine the Monkey in more detail as I explain the symbolism of 369. This will help everyone understand why I named the Monkey Nikola.

In the image, the Monkey's tail forms a spiral centred over the Great Pyramid. This design suggests that the electromagnetic field (EMF) radio frequencies captured by the Great Pyramid are mitigated and directed upwards to the top of the pyramid, with each energy stream becoming quantumly entangled. The spiral induces a natural counterclockwise rotation, forming the beginnings of a low-pressure cell referred to as the "SEED" in a vortex.

When we examine the circular rings of the Monkey's tail, we notice four rings rather than the expected three. This is correct for all vortex functions. In Vortex Dynamics, the first rotation, or the innermost ring of a vortex, is considered the "Seed."

Just as a seed transforms and disappears as a plant matures, absorbed into the tree's life force, the ignition to initiate the rotation is designed to capture radio frequencies and encourage them to rotate counterclockwise in a low-pressure direction.

The Seed represents the birth of the vortex and has no purpose other than to initiate the direction of rotation.

THE GIZA PARK PROTOCOL HOW TO START THE SUN

If we count the other rings of the Monkey's tail, we find three rings in total.

Now, let's look at the Monkey's feet. Each foot has three toes, adding up to six toes in total.

Turning our attention to the Monkey's hands, it becomes clear that this image has been deliberately manipulated to symbolise certain numbers. Nikola's right hand has four fingers, and his left hand has five fingers. There we go again: the number nine (9).

369 – any questions?

Through extensive forensic investigations, I have determined that the Nikola Monkey identifies energy systems. As you will see, 369 is the numerology code identifying numerous systems at the Giza Industrial Park in Egypt.

The number three at the Giza Plateau represents the Earth's energy streams: EMF, Schumann, and Telluric. It also encompasses the Osiris plasma scalar longitudinal waves, which is why Osiris is depicted in green, symbolising new birth on Earth and abundance.

We should remember that the Great Pyramid also emits powerful piezoelectric currents into the troposphere, directed towards the ionosphere, 37 miles above Earth, where it plays a protective role.

Six, when we consider the second pyramid, known as the Plasma Reflux Pyramid, represents the influence of plasma and its relationship to the Monkey's reproductive organs. Recall that human male sperm is 95% plasma.

The third pyramid, the Electric Pyramid, is hot—extremely hot and explosive, producing powerful plasma lightning discharges every second.

THE GIZA PARK PROTOCOL HOW TO START THE SUN

As we can see, I have identified the three pyramids and the Osiris Tomb as being designed and arranged systematically to generate plasma.

When I think of plasma now, I think of Nikola and 369.

With that in mind, I have identified the numerology of 369, which corresponds to some of the many systems at Giza. You will see more compelling data as we continue.

Now that I have shared much first-time information on the Nazca Monkey and how I believe it fits into the Egyptian narrative, I would like to explain how I used the Nazca Lines as a treasure map, following it from Nazca to Egypt.

See image.

As you've already seen, I believe the Monkey's design is, without a doubt, the most bizarre and perplexing of the Nazca Lines. Just look at all those squiggly, confusing lines.

But what do they mean?

THE GIZA PARK PROTOCOL HOW TO START THE SUN

That's a great question. I've been trying to answer it for over a decade and a half, and today, I must admit that I've made some progress. I've accumulated data I'd like to share that helped me determine where the Monkey belongs on the Giza Plateau.

I can't claim to understand everything, but I do want to highlight some aspects that are undeniably connected to Egypt.

This is slide one of the Nazca descriptions. After hundreds of attempts to accurately align the Monkey, I measured every line of this intricate Nazca artwork.

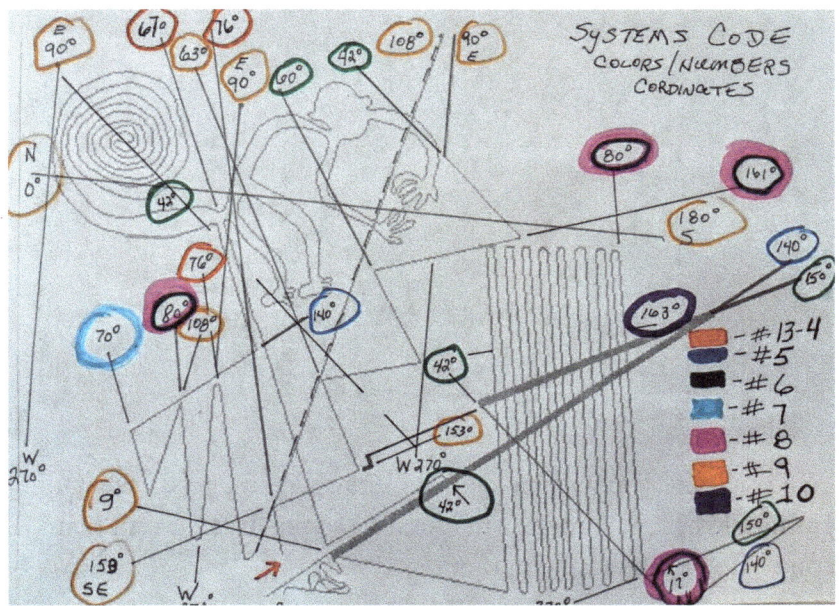

To date, I have identified nearly 100 pieces of valuable information from the Nazca Monkey, with about half of them directly corresponding to systems at the Giza Plateau.

This is the first of its kind in terms of historical information in modern times, and you will learn about all of them.

The angle of the sunrise, at 63 degrees, once again matches the far side of the smallest of the three pyramids, now known as the

THE GIZA PARK PROTOCOL HOW TO START THE SUN

Electric Pyramid, which is also set at 63 degrees. This cannot be a coincidence.

And that's not all—there's much more to uncover.

Next up is the same image with Sirius.

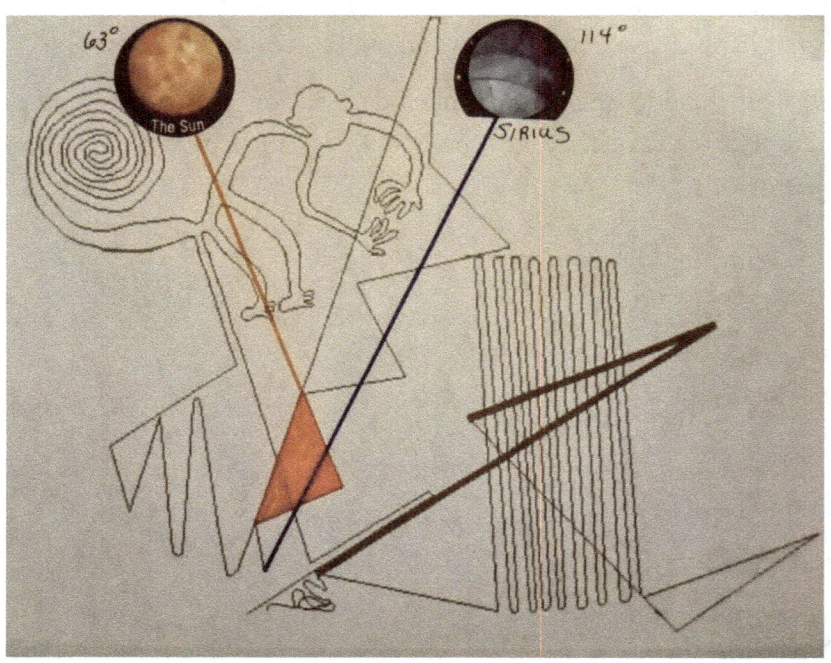

Now, take a close look at the enlarged image of the Orange Pyramid design, where the orange Sun (+) positive current connects to one side of the orange pyramid.

I'm deliberately choosing not to refer to these three pyramids on the Nazca Monkey by their designated names for now. Instead, they will be called the green, red, and orange pyramids. Soon, it will be clear to you why I've made this choice.

Like a switch, Sirius's blue (-) negative current connects to the starting circuit and becomes energised.

THE GIZA PARK PROTOCOL HOW TO START THE SUN

Notice that Sirius is at an angle of 114 degrees. I drew the blue line outward at that angle to observe where it led.

Next, take a look at the enlarged image of the Nazca line at the start and stop points. Two starts and two stops create two separate images within the same canvas.

THE GIZA PARK PROTOCOL HOW TO START THE SUN

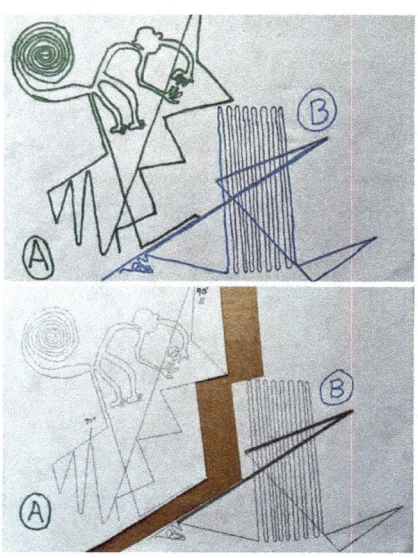

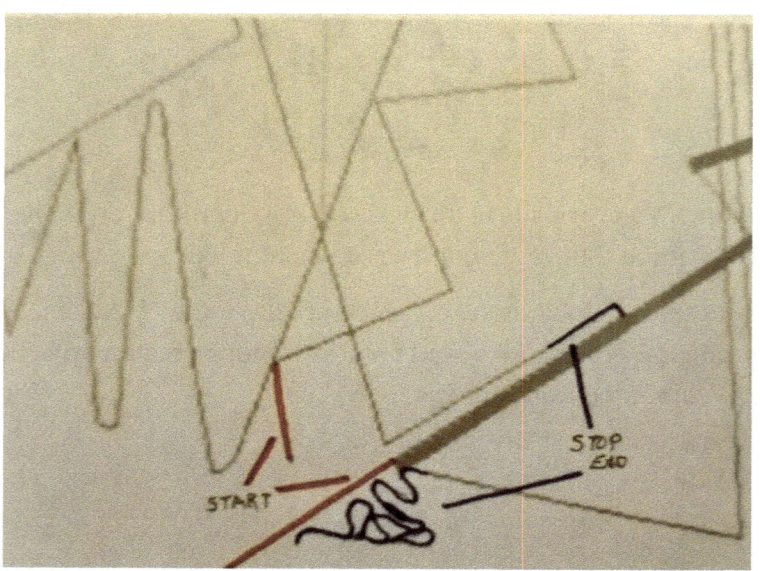

The point I want to draw your attention to is the red start, the one pointing upwards to the left. The red line marks the beginning of the Monkey drawing, and it is precisely where Sirius energy is targeted at the Helical Rise of Sirius on the Summer Solstice.

Coincidence? Nope.

THE GIZA PARK PROTOCOL HOW TO START THE SUN

Also, where the Sun and Sirius intersect is where you'll find the Crook and Flail of the Pharaohs—just like the symbol of Elon Musk's X.

Now, take a look at the enlarged image I call "Ignition."

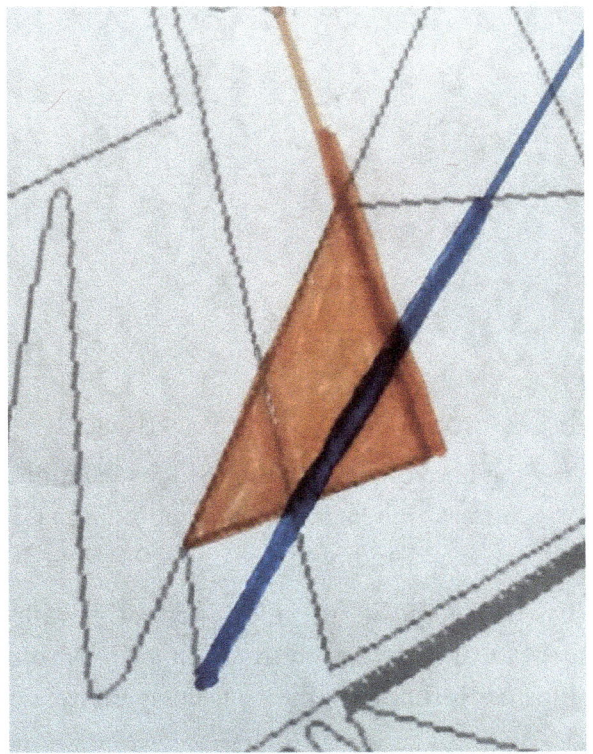

So far, I have offered the Sun and Sirius alignments on the Nazca Monkey at the Giza Plateau on the Summer Solstice 7,000 BC.

We are off to a Great start.

THE GIZA PARK PROTOCOL HOW TO START THE SUN

In the drawing above, you can see the two locations of the Sun during the 1st Cycle, which lasts 3 hours, beginning at 6 AM with a sunrise at 63 degrees and ending precisely at 9 AM at 108 degrees.

These two solar positions align perfectly with the Great Pyramid Causeway and the area in front of the Giza Sphinx, located at the Ionic Plasma Reflux Pyramid Causeway on the Giza Plateau.

Are you convinced yet?

The impressive sacred number connecting the Sun and the Causeway is 108 degrees, which marks the end of the first three hours of solar intervention.

Now, look at the next image of the Nazca Monkey. I measured every line on the glyph and discovered so many alignments that they can't possibly be coincidental. Below, I will list some numbers that draw attention to familiar figures I identified at the Giza Plateau, long before I discovered the relationship between Nazca and Egypt.

THE GIZA PARK PROTOCOL HOW TO START THE SUN

As you can see, the Nazca Monkey contains a wealth of information. Once, I laid out the Nazca Monkey glyph according to the Cardinal Points of the Earth's N/S/E/W coordinates at Giza.

This precision revealed the exact alignment of the Nazca Monkey glyph coordinates, leading me to uncover a remarkable symmetry between the physical Giza Industrial Park and the Nazca Monkey glyph.

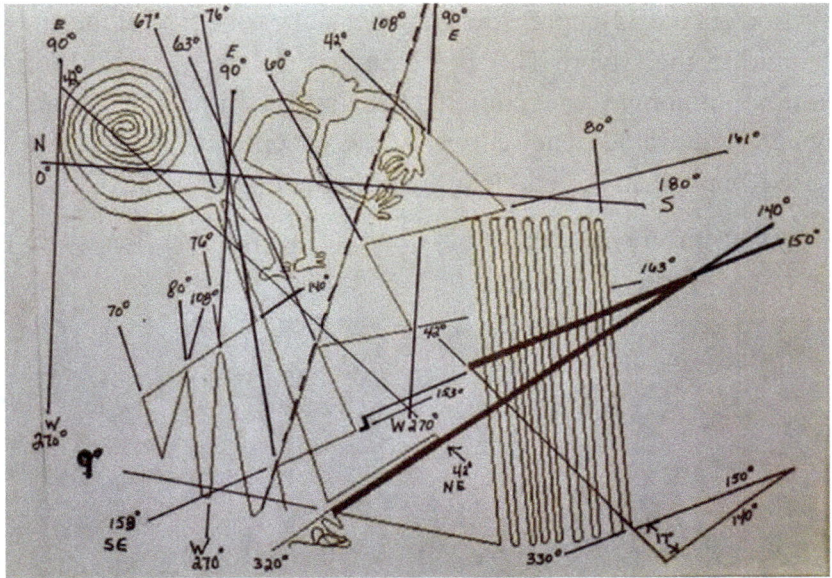

A 90-degree picture-framed Monkey with two locations, and I added the third at the tail to accurately locate the Nazca Lines.

The first line I want to mention is the angle of the Sun at the Helical Sunrise on the Summer Solstice, as the Sun traces its path across the sky at the Tropic of Cancer. This measurement is always 63 degrees and remains the same around the globe on that one day of the year when the planets and designed systems are in perfect alignment.

If you notice the small pyramid at the bottom of the glyph, its base is set at a 63-degree angle, aligning with the Sunrise at 63 degrees. Another coincidence?

THE GIZA PARK PROTOCOL HOW TO START THE SUN

This provides unequivocal evidence of the extraordinary relationship between the Orange Pyramid and the Summer Solstice Sunrise.

Also, notice the 108-degree line along the bottom of the three pyramids on the glyph. 108 is the precise angle of the Causeway, which runs past the Sphinx from the Ionic Plasma Reflux Pyramid. The 108-degree angle appears twice on the Monkey Glyph.

Look at the 42-degree line, which matches one side of the Green Pyramid in the Glyph. This is the angle the Great Pyramid forms with the Rainbow Connection. I've included an image from one of my late great friends, Rich Jarvis, whose exceptional work is greatly missed. He had an incredible eye for shapes and alignments.

As you can see, there's a second 42-degree line, the angle of one of the Stop/End lines.

THE GIZA PARK PROTOCOL HOW TO START THE SUN

As I have stated, I have documented two three-hour periods of operation for the Egypt Weather and Climate Modifications System. The first solar alignment provided by Nazca is the sunrise at 63 degrees, which appears once. The second alignment I mentioned above is 108 degrees, shown twice by Nazca. The Sun aligns with the angle of the Causeway coming from the second pyramid, the Ionic Plasma Reflux Pyramid.

The third location of the Sun marks the end of the second three-hour solar influence, which happens to correspond with the opposite side angle of the Electric Pyramid, at 153 degrees. To do the math: the Sunrise is at 63 degrees, and the Sun travels 15 degrees per hour. Therefore, in three hours, 63 + 45 degrees equals 108 degrees. Adding another 45 degrees gives us 153 degrees.

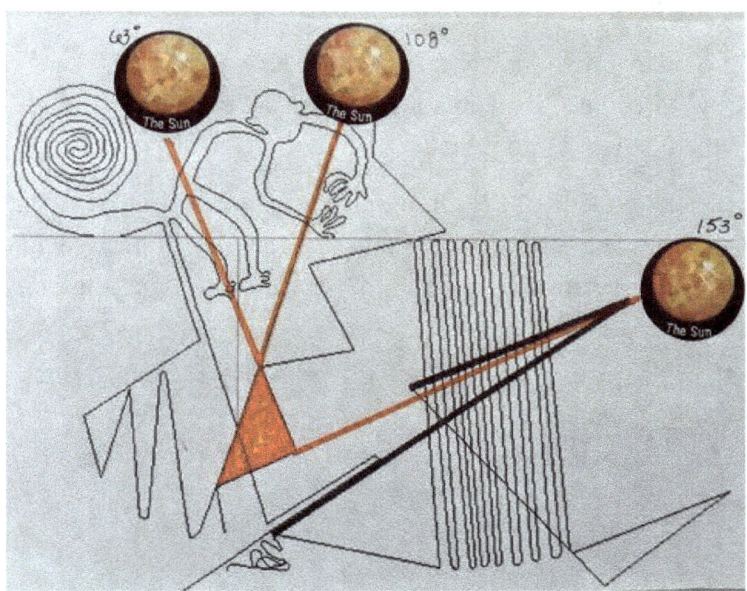

This image provides sufficient and accurate information that correlates the Nazca Monkey Glyph from Peru with the 6-hour solar cycle at the Giza Plateau, without any justifiable argument.

Before I shift gears with additional data to support my work, I also want to mention how the 114-degree angle of the Helical Rise

THE GIZA PARK PROTOCOL HOW TO START THE SUN

of Sirius on the Summer Solstice connects to the start line used to draw the Monkey on the Nazca Lines.

You can't make this up; this is not a coincidence. Whoever placed the Nazca Monkey Glyph in Peru had extensive local knowledge of the Giza Plateau in Egypt, as well as a deep understanding of many sciences, including Astronomy and Earth Cycles.

Now, I've provided everyone with a brief list to collaborate on the design and layout of the Nazca Monkey in Peru, using the four cardinal directions from both Nazca and Egypt to confirm the accuracy.

I've reflected the Orange Pyramid's angle with the Sunrise's precise angle of 63 degrees—that's five alignments.

I've also included two 108-degree correlations that align perfectly with the Giza Causeway from the second Pyramid—this brings us to seven alignments.

The second side of the Orange Pyramid is the 153-degree angle. At noon, the Sun was directly above the Giza Plateau, when the Egyptians worshipped Horus, the Sun God—making it eight alignments.

That in itself is amazing.

Are you convinced?

THE GIZA PARK PROTOCOL HOW TO START THE SUN

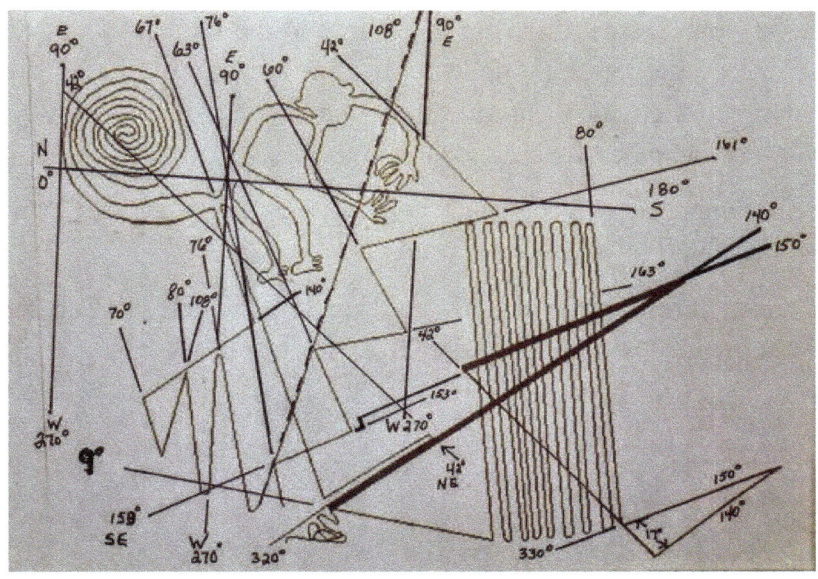

Ninety degrees is due East and is represented perfectly by the Lioness Geyser Sphinx.

THE GIZA PARK PROTOCOL HOW TO START THE SUN

As you can see, there are three lines to the East at 90 degrees. The one on the left is a duplication of the Nazca Lines' East/West 90-degree line, which I used in the same way a surveyor would, to help me locate the Monkey, ideally at the Giza Plateau.

I framed the Monkey with vertical lines at 90 degrees, which matched "Nikola" perfectly.

Sixty-three degrees is the exact angle at which the Sun rises during the Helical Sunrise at the Giza Plateau, dating back to 7,000 BC.

THE GIZA PARK PROTOCOL HOW TO START THE SUN

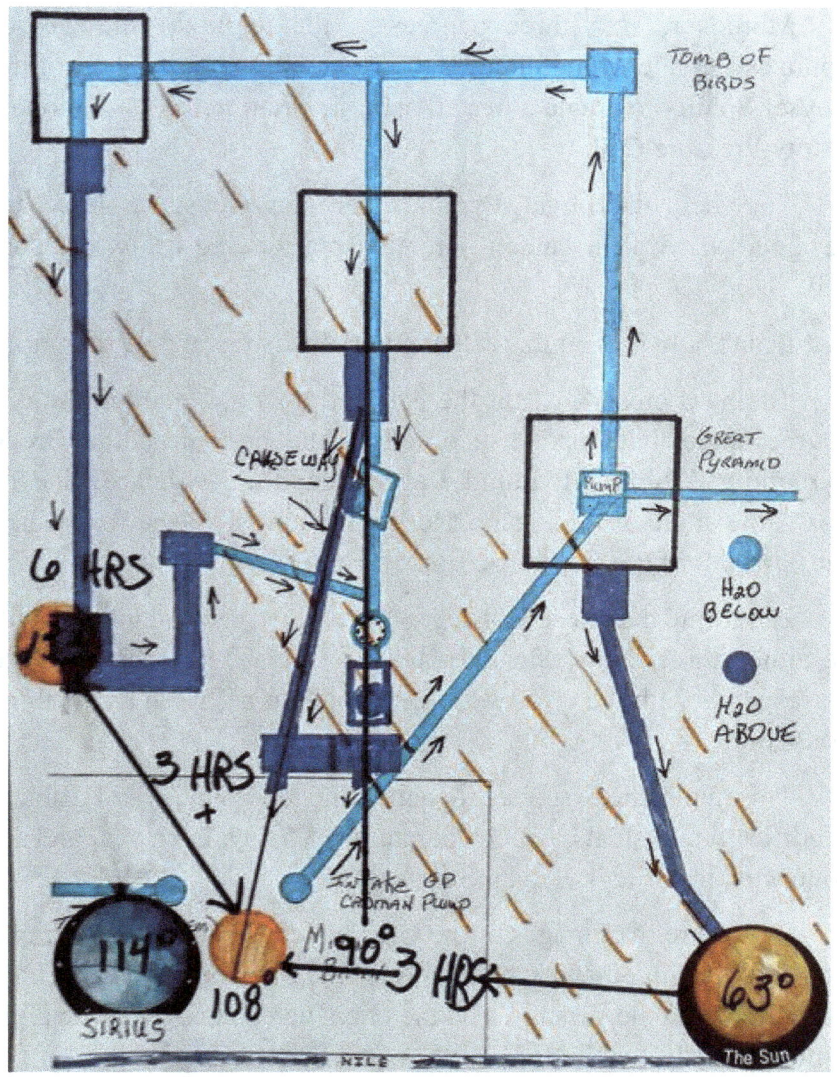

More collaboration is needed with the data from the Giza Plateau, which aligns with the coordinates provided by Peru.

Sixty-three degrees is also the angle of the final section of the Causeway from the Great Pyramid, where it directs its unused, structured, treated, and plasma-infused waters into the Nile. This allows the waters to continue on their journey to the Mediterranean Sea.

THE GIZA PARK PROTOCOL HOW TO START THE SUN

Moreover, sixty-three degrees is the angle of the Geyser Sphinx's Temple, which focuses the Helical Sun on the chest of the Geyser Sphinx to create a heated area, facilitating the formation of a Low-Pressure Cell.

Together, the Great Pyramid and the Geyser Sphinx are designed to work in tandem, creating plasma and a low-pressure cell.

I can't help but laugh, realising that this is just the beginning.

Taking a closer look at the Nazca drawing, I marked the 63-degree angle that appears to intersect the bottom of the Green Pyramid and the Red Pyramid. I also added the 108-degree line, as this is the angle of the Plasma Reflux Pyramids' Causeway, which runs in front of the Geyser Sphinx.

Is it ironic that these 108 degrees are exactly where the Sun will be aimed three hours after the Helical Sunrise? This line in Peru corresponds with the Causeway when the Sun has risen and is three hours old.

This marks the first phase of activating the Weather and Climate Modification Systems, which operate for 295 days each year and are connected to Sirius's Helical Sunrise.

Below are two images of the Giza Plateau. The first is a simple drawing detailing the exact locations of the Sun and Sirius, and how they align with the structures on the Giza Plateau and the Nazca line simultaneously for the five cycles of the Diurnal Operations of the Egyptian Weather and Climate Modification Systems.

THE GIZA PARK PROTOCOL HOW TO START THE SUN

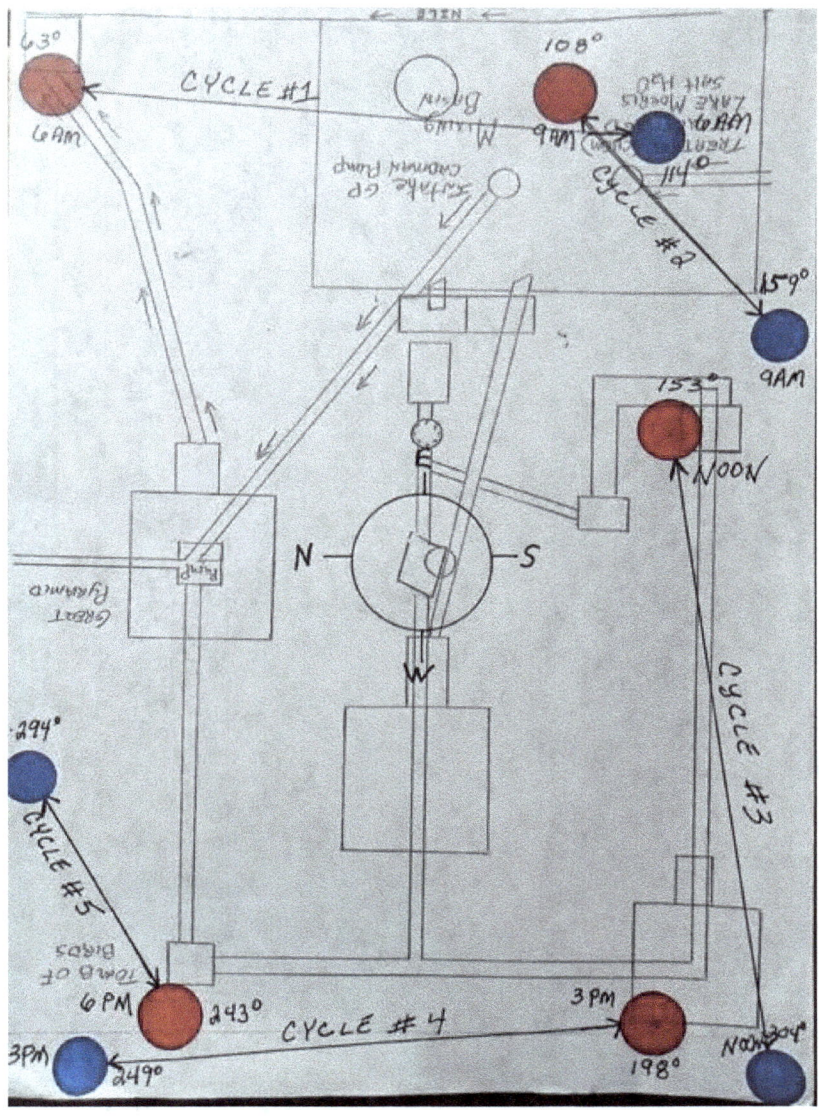

In the second image, I included "Nikola," the Nazca Monkey, at his designed location, showing us the two most important places, both pyramids, identifying the Great Pyramid with his tail and the Electric Pyramid with his hands.

I will get to his legs shortly.

THE GIZA PARK PROTOCOL HOW TO START THE SUN

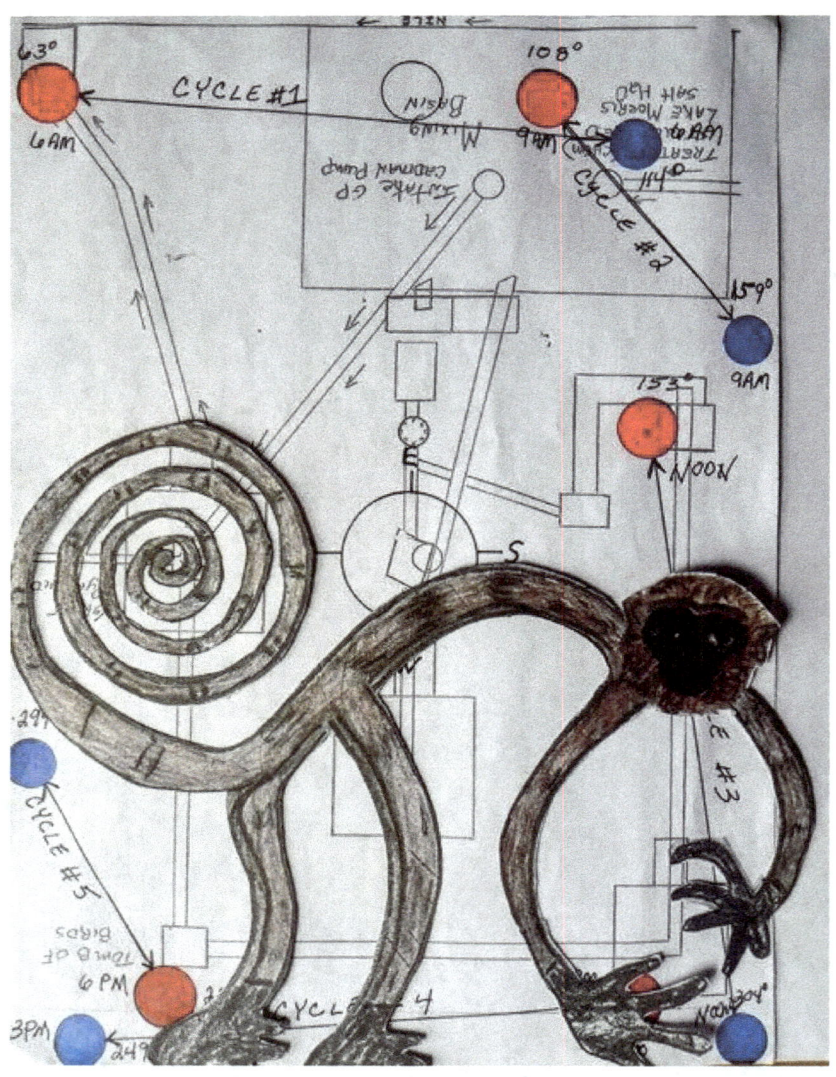

Now, it would be appropriate to discuss the second component of the Nazca Line, Segment Blue B, which I have determined is the system's operation time cycle.

THE GIZA PARK PROTOCOL HOW TO START THE SUN

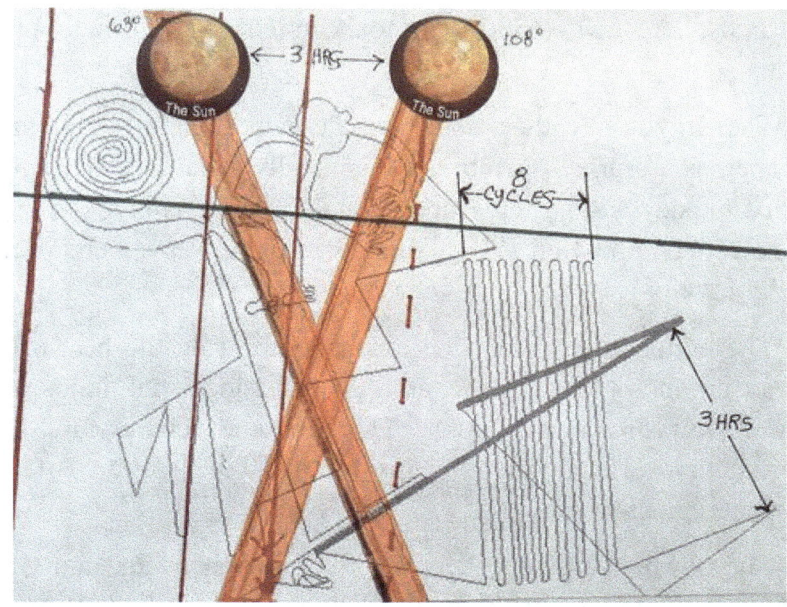

In Segment B, we are observing a method to detail the time cycle of operations, which illustrates how the system is intended to function.

As I mentioned earlier, this image represents the first 3-hour cycle of five. The peculiar, organised radiator-like arrangement features eight identical tubes, each shaped precisely the same. The darker straight lines across the radiator are visually connected and include the Start/Stop switch, as well as all eight tubes.

I believe this reflects one cycle within a three-hour period, as demonstrated. I am calculating 3 hours multiplied by 8 cycles, totalling 24 hours, which equals one day and one night—essentially one full rotation of the Earth.

While the five cycles, adding up to 15 hours, represent the heat cycle of the daytime, the systems are influenced by the drop in temperature at night, which shifts their focus towards encouraging rain. This is because the pumping system never stops sending plasma-treated water through the top of the Geyser Sphinx head into the troposphere, up to 6,500 feet into the atmosphere, throughout the night.

The Egypt Weather and Climate Modification System is designed as a diurnal operating system, fully functional during the heat of the day and partially operational at night. This operation is influenced by the Sun's local impact on the ionosphere during daylight hours.

The temperature inside the ionosphere during the heat of the day is around 440°F, while at night, the localized ionosphere environment drops to -200°F. This extreme temperature shift significantly affects the expansion and contraction of the ionospheric membrane.

The physical distance between the ionosphere and the Great Pyramid is 37 miles during the day heat cycle and 54 miles during the night cold cycle. I believe this natural hot and cold cycle serves as the activation mechanism, the "on/off" switch, that triggers the Great Pyramid's powerful lightning discharges, which agitate the core of the ionosphere.

Do you recall the Aurora Borealis being visible at unusually low altitudes in 2024? This was a direct result of the Coronal Mass Ejections (CMEs) from the Sun.

My research, based on NASA's ionosphere scientific data and their latest findings on recent CMEs, has revealed a significant impact on the ionosphere. Scientists have stated that each CME from the Sun increases the positive (+) plasma ions within the ionosphere, which leads to a substantial rise in the ionosphere's interior temperature and an expansion of its circumference.

The increase in positive (+) plasma ions is why we see the beautiful rainbow colours in the sky at lower latitudes and experience disruptions in electrical communications. When the blast of intense piezoelectricity from the Great Pyramid enters the ionosphere, similar reactions occur, resembling the effects of solar CME discharges.

It seems logical that, when the day cycle is active at Giza and the ionosphere is 37 miles away, the range of the piezoelectric discharge penetrates the ionospheric shell. This causes a reaction of attraction between the residual plasma inside the ionosphere.

These excited plasma ions, with temperatures of 440°F and a positive (+) current, are attracted to the Great Pyramid's piezoelectric (-) negative current. This natural attraction causes many of these plasma (+) ions to cling to the piezoelectric discharge from the Great Pyramid and are subsequently exported from the ionosphere.

The image below depicts an electric pyramid with a lightning strike, which I have included to offer a visual representation of the function of the Great Pyramid's piezoelectric plasma discharge.

THE GIZA PARK PROTOCOL HOW TO START THE SUN

Imagine being 100 feet beneath the ground, with the Cadman Pump located below the Great Pyramid (GP), making a mechanical stroke every second and generating 3,200 lbs of pressure, which echoes through the Great Pyramid's resonating and vibrating chambers.

Consider the vibration of this mechanical force resonating linearly and interacting mechanically with the electrical states in the crystalline materials of the granite. I will later detail the electromechanical interaction of piezoelectricity, but for now, it is important to grasp the scale and magnitude of this ancient technology.

The steering and collection of these superheated plasma ions are among the functions designed for the second pyramid, the Ionic Plasma Reflux Pyramid, which will be discussed in further detail.

The next chapter will cover the Ignition Protocol—"How to Start the Sun."

Now that I have described the two images on the Nazca Monkey Glyph, you will be able to see how the diurnally-designed system operates in its cycles. In other words, during the summer solstice, the pyramid systems are activated at the helical rise of Sirius at sunrise, and the rain begins. The system operates for five cycles, each lasting 3 hours, a total of 15 hours. This generates heat and humidity during the day and produces copious amounts of rain due to the drop in ionospheric temperatures at night. This is how the Sahara was once green.

The Sahara Desert, once abundant in life—how was this possible? The simple answer is plasma, the most abundant ingredient in our universe. Where there is plasma, there is birth and abundance. Thanks to the Egypt Weather and Climate Modification Systems.

For now, I hope you have a better understanding of how these ancient systems were designed in such an economical and environmentally sustainable way. Absolute genius. I am thrilled to finally share these magnificent discoveries.

I feel confident that I have provided sufficient collaborative evidence for most of you to read my research and view things from a different perspective.

The next chapter will include an actual "Start-up Procedure" for the Egypt Weather and Climate Modification System. For the first time in modern history, you will begin to recognise the design and function of pyramid systems on Earth.

Chapter Ten:
The Egypt Taurus Correlation Theory

The Taurian Astro-Blueprint for the Egyptian Pyramid Systems

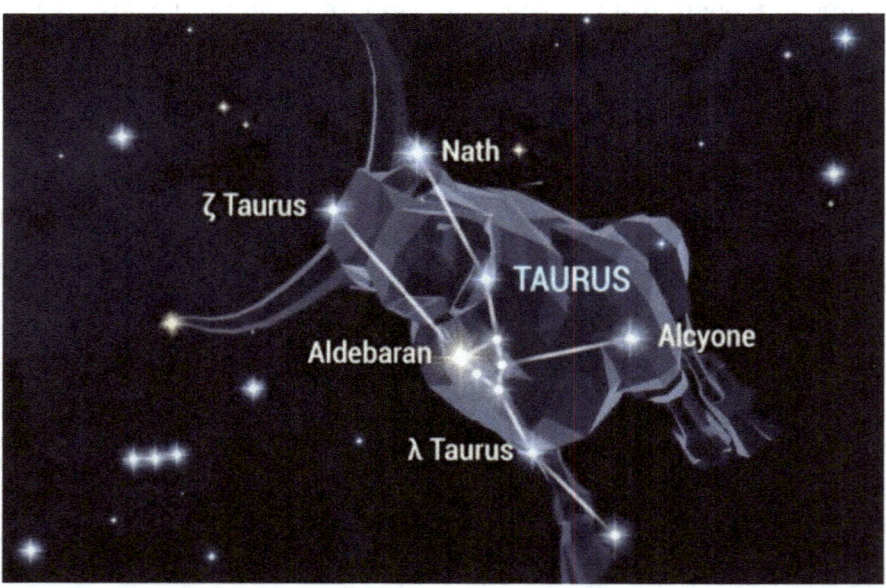

One of my favourite discoveries in Egypt was realising the apparent close relationship between the designers and builders and the cosmos. This connection was evident in many ways, including their understanding of astronomy, planets, Sirius, the Sun, and the Heliacal Sunrise cycles.

Most people are familiar with Robert Bauval's extraordinary "Orion Correlation Theory." While working in Egypt as a surveyor, Bauval observed how closely three stars in the Orion constellation—

specifically Orion's Belt—aligned with the layout of the three largest pyramids on the Giza Plateau.

As a nuclear, industrial, mechanical, and marine systems analyst, as well as a professional in production, research and development, and start-up planning and engineering, I have the flexibility to observe and recognise the complexities of many so-called Old Kingdom Egyptian systems. My approach involves either building upon the work of past researchers or conducting deep research, employing reverse engineering and forensic investigative methods to determine the function and purpose of each system.

Over the past 15 years, I have dedicated myself to uncovering and analysing this information, and I am eager to share my findings.

Before moving on to the next chapter, where I will provide further details on these systems, I want to describe an accidental discovery I recently made. I have concluded that Egypt's geography and many of its ancient systems are vast and extraordinary.

To begin, I would like to introduce the Taurus Correlation Theory by providing a brief history of the significant relationship between the ancient Egyptians and the bull or cow, which they were not alone in revering.

Bull cults were widespread in Egypt from the First Dynasty (Early Dynastic Period). The powerful and virile bull was associated with the pharaoh, who sometimes took the epithet "strong bull of his mother." The bull's mother was venerated in numerous locations across Egypt, including Memphis, where Hathor—one of the principal goddesses of the Egyptian pantheon—was depicted as a cow.

During our travels, Heidi and I visited the beautiful Hathor Temple in Dendera, which houses the Temple of Mammisi, also known as the "Birth House of Horus."

THE GIZA PARK PROTOCOL HOW TO START THE SUN

The great Greek historian Herodotus, to whom Egyptians owe much for documenting their oral histories spanning over 3,000 years, provides valuable insights into these ancient traditions. However, when discussing early Egyptian history, there is considerable room for error and speculation.

The Egyptians meticulously recorded that the most famous and sacred bull was a particular type known as the Apis Bull. These bulls were believed to be divine, identified by a distinctive diamond-shaped white mark on their black heads. Treated as living gods, they were groomed, fed the finest foods, housed in special quarters, and worshipped daily.

THE GIZA PARK PROTOCOL HOW TO START THE SUN

Here is one I patted on the head in India.

When a bull died, it was mummified and buried with full royal honours.

Many archaeologists believe that the massive sarcophagi at the Serapeum were used to bury the Apis Bulls. However, I find this claim ridiculous—just one of many absurd theories that resemble poorly written movie plots, a common occurrence when discussing Egypt.

One particularly fascinating story from Herodotus tells of the Persian King Cambyses, who had conquered Egypt and was ruling at the time. He witnessed the Egyptians celebrating the Apis Bull, a festival that lasted seven days. During this time, the sacred bull was led through the streets as people cheered.

According to legend, any child who inhaled the breath of the Apis Bull would gain the ability to see the future and become a powerful oracle.

Annoyed that the Egyptians were celebrating something other than himself—their conqueror—Cambyses ordered his soldiers to find this so-called "god" and bring it to him. Once they captured the Apis Bull, Cambyses killed it, had it cooked, and publicly ate it, horrifying the already defeated Egyptians.

As I mentioned, the Apis Bull Cult was the most famous in Egypt, but two other prominent bull cults also existed: the Buchis Bull Cult and the Mnevis Bull Cult.

Beyond Egypt, the reverence for bulls extended to other cultures. One well-known community, the Canaanites, worshipped a bull they called "El" as their supreme god.

From the very beginnings of Egyptian civilisation, the bull was sacred and widely worshipped.

I firmly believe that my identification of the so-called Old Kingdom systems in Egypt is correct—they were constructed based on a mirrored image of the Taurus constellation imprinted upon the sands of the Sahara Desert.

This is where I began to answer the question: why?

My discoveries led me on a journey across the desert, spanning precisely 144 miles from the city of Alexandria on the Mediterranean Sea to the Pyramid of Amenemhat III at Hawara. Additionally, the path extends a few extra miles across the Mediterranean Sea to one of the most powerful Earth vortex points—designated as UVG1—just off the coast of Port Said, near the entrance to the Suez Canal, which was constructed by France in 1897.

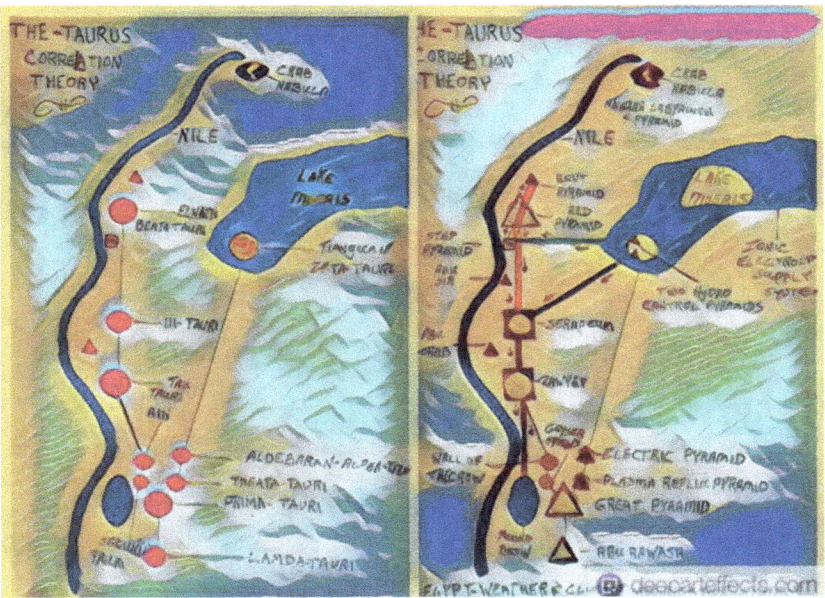

Orion's magnificent constellation consists of ten officially named main stars and a nebula. However, only the three stars in Orion's Belt are directly correlated to three structures on Earth.

Following my identifications, I plotted the GPS coordinates of each system. I then overlaid a mirrored image of the Taurus constellation, scaling it to align with the thousands of years old ancient industrial systems.

The match was astonishingly precise, too accurate to be mere coincidence. The relationship was immediately evident.

THE GIZA PARK PROTOCOL HOW TO START THE SUN

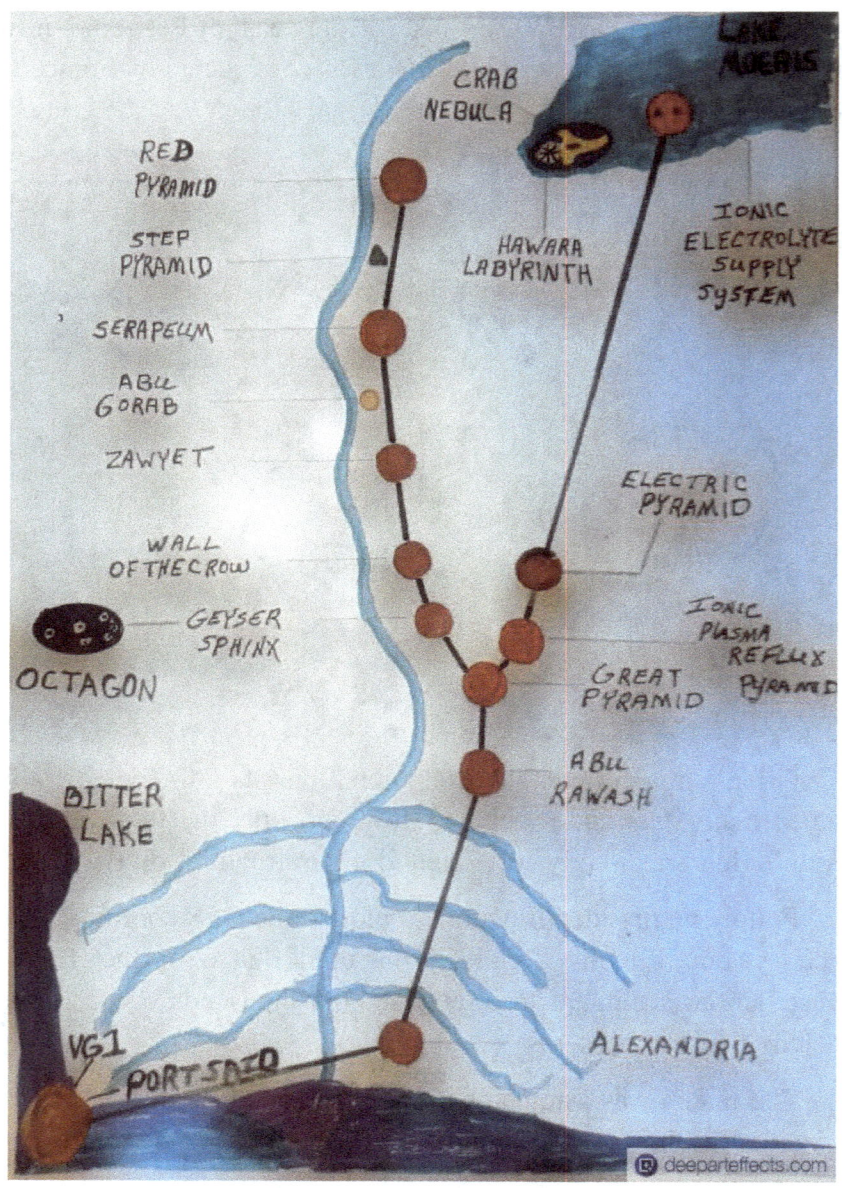

The Taurus constellation has 17 officially named stars and is a highly noticeable and vast constellation in the night sky. It also contains the Crab Nebula and is paired with the Pleiades, known as the Seven Sisters, a name given by the Greeks, or identified as the "Leg of the Bull" by Wayne Herschel.

THE GIZA PARK PROTOCOL HOW TO START THE SUN

My research has identified 12 stars, the Crab Nebula, and the most powerful Earth vortex, all of which align with these ancient Earth systems, bringing the total to 14. The location of the "Pleiades," detailed below, adds a 15th system.

It took longer to complete this identification because the final piece of the puzzle was missing: pinpointing the location of the Pleiades system, or Alcyone, since it is so closely associated with Taurus, particularly in terms of energy exchange. However, I needed to first identify the other systems to establish an accurate reference point before beginning my search in the east.

I must address the absence of the Pleiades star cluster, known as the Seven Sisters or, in ancient times, as Alcyone, from this treatise. Observing the Taurus constellation in the sky, I could not determine with certainty the exact distance at which the Pleiades would appear on land or what I might discover in that location.

Once I identified which Taurus stars aligned with the Egyptian systems on land, I became convinced that the Pleiades would be located directly east of the Giza Sphinx.

Using Google Earth, I measured precisely from the Giza Sphinx toward the east, following that heading with curiosity to see what I might uncover. I determined my heading to be exactly 90 degrees east. After traveling mostly over desert terrain, 33 miles from the Giza Sphinx, I made a remarkable discovery—a tangible, physical representation of the modern-day Pleiades.

I arrived at the perimeter of a vast complex consisting of eight substantial buildings arranged in a perfect octagonal configuration. Each building houses the headquarters of one of the eight branches of the Egyptian Armed Forces within the Egyptian Ministry of Defense.

THE GIZA PARK PROTOCOL HOW TO START THE SUN

The eight-sided complex is called the Octagon. The buildings are arranged in a large circle, with two central monumental buildings at the centre. There are ten buildings in total.

If we take a few minutes to discuss the physical characteristics of the Pleiades, we know it is sometimes called the Seven Sisters, a cluster of stars. We often believe it consists of only seven stars, whereas the Pleiades Cluster is actually made up of over 1,000 stars.

Check out the image of the Egyptian version of the USA's Pentagon. It is the largest military complex in the world. Could this represent the actual Pleiades in the sky? I certainly think so.

I immediately noticed that the eight-sided Octagon also mimics the eight-sided Great Pyramid.

While on the topic of the Pleiades, I find Wayne Hershel's theories about Abu Sir fascinating. He presents a compelling argument that Abu Sir symbolises the Egyptian ritual known as The Leg of the Bull, another symbolic association with the Pleiades.

My only issue, which is not part of Wayne's theory, concerns the location of Abu Sir. Abu Sir is within walking distance of Abu Gorab, three-quarters of a mile to the north. The Sphinx is 6.5 miles further north, and the Step Pyramid is only two miles south.

THE GIZA PARK PROTOCOL HOW TO START THE SUN

Compared to the recent Octagon, the location itself does not align with where it would be on Earth in relation to the Pleiades.

In my opinion, the Octagon closely represents the Pleiades, if not exactly, for the following reasons:

According to my Egypt-Taurus Correlation Theory, the Pleiades would be due east from the Giza Sphinx. I chose this direction to look for either ancient or modern symbolism.

The Octagon is a magnificent example of modern sacred geometry, symbolising the Pleiades. It represents power and might as the military command centre for Egypt.

The Octagon parallels the US Pentagon in Washington, DC.

It is significant that the Octagon has eight sides, just like the Great Pyramid. It functions as a massive antenna, designed to capture powerful electromagnetic fields in this ideally located system.

The Ptolemaic Dynasty, the 33rd dynasty based in Alexandria, was the most powerful in Egyptian history.

The number 33 is significant, representing the highest degree in the Ancient and Accepted Scottish Rite of Freemasonry.

The House of the Temple, home to the Supreme Council (33rd degree) of Freemasonry in Washington, DC, has 33 outer columns, each 33 feet tall.

The Wall of the Crow, part of the Giza Park Industrial Systems, has intriguing connections:

Height: 33 feet

Width: 33 feet

Length: 650 feet (which reduces to the number 11)

THE GIZA PARK PROTOCOL HOW TO START THE SUN

It is directly linked to the giant orange star Epsilon Tauri, known as AIN in ancient Egypt, which represents the right eye of the bull in the Taurus constellation.

The left eye of the bull is associated with the Electric Pyramid, corresponding to the star Alpha Tauri or Aldebaran, known as the "Bull's Eye."

The exact distance from the Giza Sphinx to the Octagon is 33 miles. Is this merely a coincidence, or a deliberate clue for further investigation?

Interestingly, I have completed the Bosnian Pleiades Correlation Theory and determined that the distance from the Sphinx to the Bosnian Pyramid of the Sun is 1,200 miles, a number associated with three in numerology. Another coincidence?

During my nine-week stay in Egypt, which included the Islamic holy month of Ramadan, I observed how deeply religious the people were. Similarly, during my 2.5-month stay in Turkey, I made friends and was invited into their homes. I noticed that the faithful in both countries pray five times daily, often in groupings of 11, 33, and 99, reflecting the "99 Names of Allah." In Islam, 33 angels are said to carry man's praise to heaven.

Historically, Alexander the Great conquered Egypt and established the Ptolemaic Dynasty, the 33rd and most powerful dynasty. If the Egyptian government chose the Octagon's location based on these numerological and historical connections, I commend them for their thoughtful planning.

Now, regarding Abu Sir's location: we have credible documentation that the Heb Sed Festival was celebrated for over 1,000 years. Abu Sir was built by Neuserra in the 5th Dynasty, long after King Den held the first festival in the 1st Dynasty. Hieroglyphs depict Den running along boundary markers during the festival to demonstrate his strength and endurance.

THE GIZA PARK PROTOCOL HOW TO START THE SUN

Many pharaohs constructed temples to participate in this sacred ritual, displaying their wealth and power. The first was at Saqqara, and in the 3rd Dynasty, the Djoser Complex was built to celebrate the Heb Sed Festival, also known as the Feast of the Tail.

I believe Abu Sir was built more recently than the Giza Pyramids and other Heb Sed locations. The pyramids at Abu Sir reflect a later, lower-quality brick construction method that has deteriorated significantly over time. Some archaeologists argue that the Red Pyramid and Abu Sir are older than the Great Pyramid, but physical evidence suggests otherwise.

Abydos and the Osirion, dedicated to the Isis and Osiris cults, were built in the 1st Dynasty and remained in use until the 13th Dynasty. Seti I later restored them and included them in the Kings List in the 5th Dynasty. Pepi I also constructed Dendera in the 4th Dynasty (2613–2494 BC). Compared to these, Abu Sir is a more recent structure.

Furthermore, Abu Sir's location does not accurately reflect the Pleiades' position in the sky. However, it does serve as a significant worship site, particularly given its proximity to Abu Gorab, a Sun Temple.

Despite the many references to the Heb Sed Festival throughout Egyptian history, the most detailed records come from Abu Gorab. Information from Amenhotep III's reign and other historical documents were discovered there, dating back to the 5th Dynasty. Abu Gorab and Abu Sir are only a three-quarter-mile walk apart through the Sahara Desert—an experience my companion Heidi and I personally undertook, with her even walking barefoot in the warm sand.

Returning to my Taurus Correlation Theory, I have identified several important sites directly east of the Giza Sphinx. This theory aligns 15 celestial markers with structures on the Sahara Desert's surface in Egypt.

THE GIZA PARK PROTOCOL HOW TO START THE SUN

Not only do the three stars of the Taurus constellation align with the three large pyramids at Giza, but the Sphinx and the Wall of the Crow add to the count, making five alignments at the Giza Plateau alone.

Today, I can match 15 stars of the Taurus constellation with Egyptian structures and systems, including the Crab Nebula, the Pleiades, and vortex UVG1, which is positioned at 63 degrees east at the heliacal rise of Sirius. This alignment extends from Port Said on the Mediterranean Sea inland to Alexandria and Hawara in Fayum, covering 144 miles.

In ancient times, achieving such precise alignments without advanced measuring tools would have been nearly impossible. In Chapter 6, I will provide a detailed tour using a system flowchart format, tracing each site's function from its origin.

This journey will begin at Lake Moeris and follow a primarily northward route to Alexandria, corresponding with the natural flow of the Nile. However, for clarity, I will present my Taurus Correlation Theory in reverse, from north to south, starting with Alexandria and the Giza Plateau, as these are the main attractions.

Now, let us examine the mirrored image I have produced, showing how the twelve stars of the Taurus constellation align with ancient Egyptian structures and systems, alongside the locations of the Crab Nebula and the symbolic Pleiades.

THE GIZA PARK PROTOCOL HOW TO START THE SUN

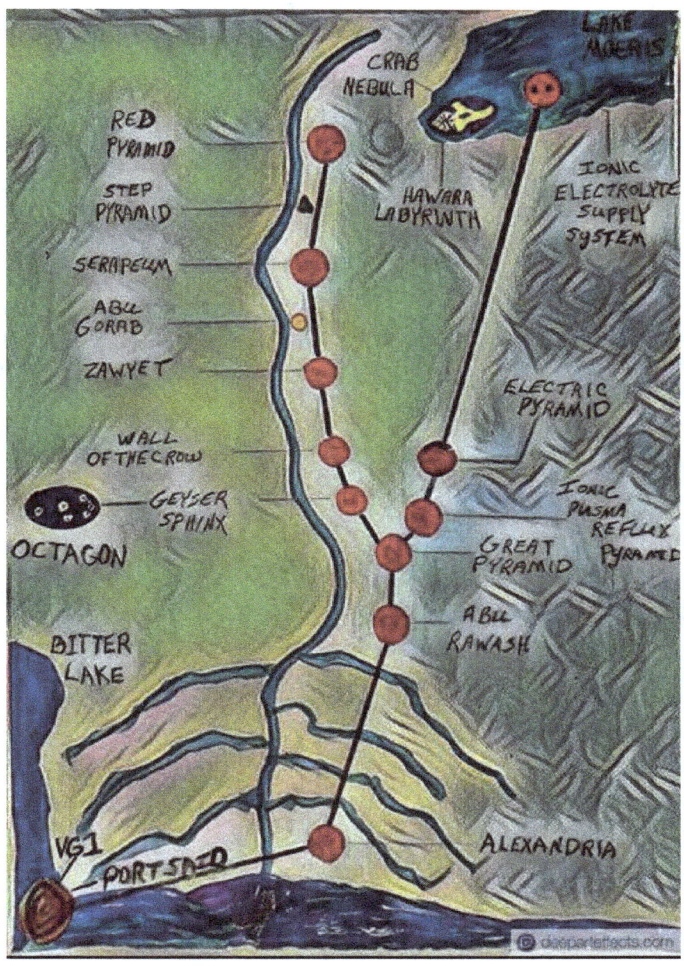

Here's a revised version with correct grammar, structure, punctuation, spelling, and improved flow:

System #1: Earth Vortex – UVG1, Mediterranean Sea (off Port Said, at the entrance to the Suez Canal)

This vortex is considered the most powerful on Earth and aligns perfectly with both the North and South magnetic lines.

Discover my fascinating find between Jerusalem and Cairo, where a perfect **Vesica Piscis** formation appears—strikingly similar to the one found in the United States' capital. In Washington, D.C., the Washington Monument stands at the centre of this geometric pattern, suggesting a high probability of the presence of the **UVG1 Earth Vortex**.

This offshore location in the Mediterranean Sea aligns 63 degrees to the east, which is the exact angle of sunrise at the summer solstice and, coincidentally, the heliacal rise of Sirius.

I find it uncanny that the capitals of Jerusalem and Giza form a perfect *Vesica Piscis*, similar to how Washington, D.C., embraces the Egyptian-inspired Washington Monument.

During my investigation, using all available information to accurately locate the UVG1 Vortex on Earth, I narrowed it down to an area offshore near Port Said. When I made the exciting discovery about the alignment between Jerusalem and Giza, I identified this location as being nearby.

My research suggests that this pattern is frequently repeated. Numerous citations reference similar alignments at the Giza Plateau, mirroring those of the Great Pyramid inland.

THE GIZA PARK PROTOCOL HOW TO START THE SUN

As our understanding of these existing systems deepens—systems that appear to have been meticulously designed by superintelligent, all-knowing beings—we may ultimately encounter a species with its own agenda.

Just off the shore of Port Said, at the entrance to the Suez Canal, lies UVG1—the most potent vortex field on the planet. This vortex is particularly intriguing due to its connection to a saltwater lake a few miles inland, known as Bitter Lake. My research indicates that this highly saline body of water, with a salt concentration exceeding 41%, is strongly influenced by the currents of the Red Sea. These currents contribute to an extremely salty seawater mixture, formed by the gradual dissolution of the salt basin in which the lake sits.

This phenomenon has a significant impact on the marine environment and the energy systems in this region of the Mediterranean Sea. The high salt content creates a dynamic electrolyte, and when the Sun rises at 63 degrees during the heliacal rising of Sirius, its light travels 272 miles from Jerusalem to the Great Sphinx at the Giza Plateau.

THE GIZA PARK PROTOCOL HOW TO START THE SUN

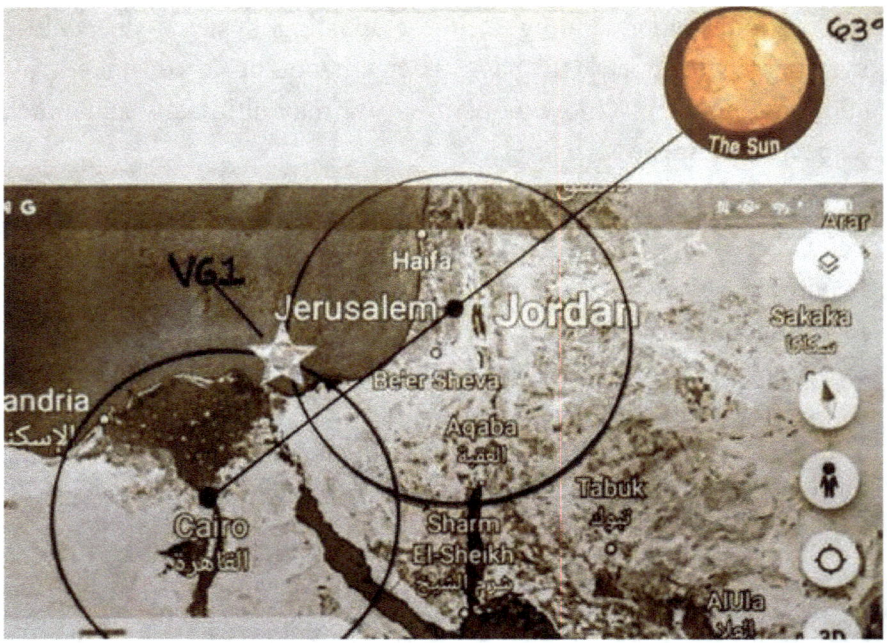

My alignments end up being a perfect Vesica Pisces relationship between Jerusalem and Cairo to an unusual extent.

On the Summer Solstice, the Sun aligns perfectly at the 63-degree angle with Jerusalem and the Geyser Sphinx, as if designed that way.

THE GIZA PARK PROTOCOL HOW TO START THE SUN

An old image is looking through the Solar Device at its target area of the Geyser Sphinx.

THE GIZA PARK PROTOCOL HOW TO START THE SUN

Perfect. Aim for this sun angle from Jerusalem to hit the Geyser Sphinx in the chest.

Here is an image of the Sun's trajectory of 63 degrees at the Giza Industrial Park.

The distance between them is 272 miles, represented numerically as 11.

Along the way lies Bitter Lake, known for its high salt content. I am trying to understand why this system serves as the primary power source for the UVG1 Vortex's energy systems.

The Taurus constellation star associated with the most powerful Earth vortex, VG1, is named Omicron. Interestingly, in Greek mythology, Omicron was a god who disapproved of people living in peace and working harmoniously to maximise their collective well-being and happiness.

THE GIZA PARK PROTOCOL HOW TO START THE SUN

It is also notable that global health leaders named the fabricated pandemic virus "Omicron."

Additionally, understanding this powerful vortex system and its location may provide insights into its influence on meteorological phenomena, such as the formation of Mediterranean cyclones, also known as "medicanes." In 2020, a Category 2 hurricane formed in the Mediterranean.

2) Alexandria

Alexandria was named after Alexander the Great of Greece, who established the last and most powerful Egyptian dynasty—the 33rd and the Ptolemaic Dynasty. At its height, Alexandria was the capital of Egypt and one of the most powerful cities in the world, second only to Rome.

From a systems management perspective, Alexandria marks the point where the Nile enters the Mediterranean Sea, a crucial mixing location. This significance will become clearer as we progress. Additionally, Alexandria lies just a short distance from UVG1, located off the Egyptian coast, north of Port Said, near the entrance to the Suez Canal.

The Taurus constellation star associated with Alexandria is Ushkaron, a powerful Tauri triple-star cluster.

The following image shows the Giza Plateau and what remains of Abu Rawash.

Abu Rawash is nicknamed the "Exploding Pyramid" because it is now just an empty crater, with every surrounding structure destroyed.

This fascinating site is perched high on a small mountaintop, approximately five miles north of the Giza Plateau. I have positively identified this location as part of a system designed as a functional component of Egypt's Weather and Climate Modification Systems.

THE GIZA PARK PROTOCOL HOW TO START THE SUN

One particularly significant system I observed in my research is Abu Rawash. Like the other three pyramids on the Giza Plateau, it had a causeway. This causeway extends down the mountainside and connects with the existing flow of the Nile River. Physical evidence strongly suggests that water once circulated through and emerged from the Abu Rawash Pyramid system.

I will provide further details on the system's function as we progress. Abu Rawash is believed to have been the final pyramid associated with Egypt's Weather and Climate Modification System.

In 2023, Heidi and I attended the *Conference on Precession and Ancient Knowledge* in Palm Springs, California, where we met Robert Edward Grant. During his lecture, he discussed his investigations at Abu Rawash. He claimed to have measured the foundation remains of smaller pyramids that had been destroyed at the site. When analysing their mathematical dimensions, he discovered a striking match to one of the symbols I had been researching on the US dollar bill.

THE GIZA PARK PROTOCOL HOW TO START THE SUN

A decade ago, I conducted an extensive search for the model used for the eye above the pyramid. It quickly became apparent that it was neither the Great Pyramid nor the other two pyramids on the Giza Plateau. Over time, I successfully identified the owner of the eye from an original painting belonging to George Washington.

However, my search for the correct pyramid only gained traction when Robert Grant pointed me in the right direction. I shook his hand and thanked him for the valuable information.

I assert that the pyramid belongs to the Abu Rawash pyramid complex, which symbolically reflects Washington on the East Coast of America as the Birth Cycle or the beginning. Meanwhile, Abu Rawash represents the West Coast and Washington as the energies enter the "Pacific Ring of Fire."

THE GIZA PARK PROTOCOL HOW TO START THE SUN

The Taurus star associated with Abu Rawash is Lambda Tauri.

Finally, we arrived at the Giza Plateau Industrial Park.

This image depicts the planetary alignments looking east from the front of the Sphinx at 6 AM in 7,000 BC, on the Summer Solstice—the day of Sirius's heliacal rise.

The following image reflects the incredible, perfect design and location of the Giza Industrial Park. The Sun, Sirius, Moon, and six planets, which make nine, all contribute their invisible quantum energies to powering up and starting the systems.

THE GIZA PARK PROTOCOL HOW TO START THE SUN

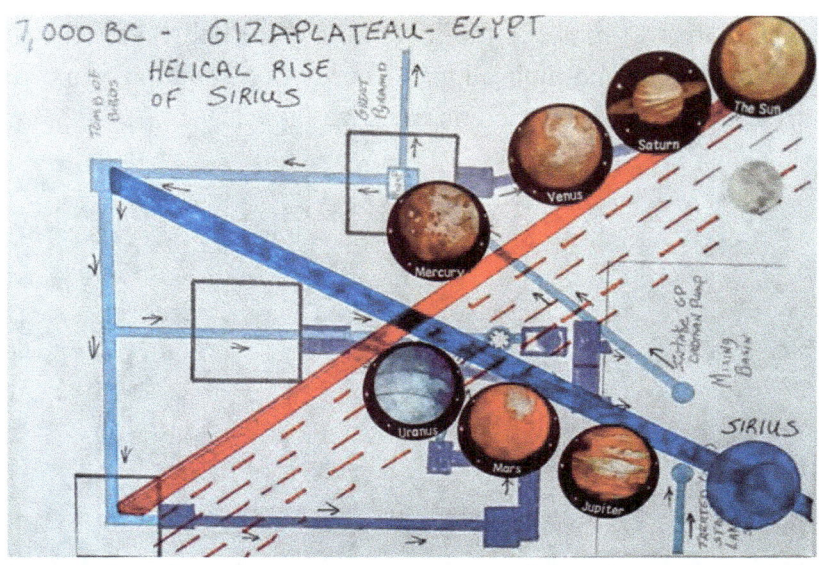

THE GIZA PARK PROTOCOL HOW TO START THE SUN

The Great Pyramid is the focal point. We will reveal the five pyramid systems and their aligned Taurus Constellation Stars; five Taurus stars are now aligned on the Giza Plateau.

The Great Pyramid is located at the very bottom center of this image. To the right is the Ionic Plasma Reflux Pyramid, and then past it is the Electric Pyramid.

Directly across from the Electric Pyramid is the Wall of the Crow, and the Geyser Sphinx is coming back towards the Great Pyramid.

THE GIZA PARK PROTOCOL HOW TO START THE SUN

I wanted to show an image of lightning emerging from the top of the Great Pyramid but had yet to succeed. However, I did find one showing lightning coming from the smaller Electric Pyramid.

4) The Great Pyramid – A Major Component of the Multisystem-Designed Weather and Climate Modification Systems

Its complex yet elegant design serves as a perfect reflection of what we are doing wrong. This example exists here, now, today.

Whenever I think of the Great Pyramid, I nickname it the *Swiss Army Knife of Pyramids* because it serves multiple functions. However, only one word truly encapsulates its essence: **PLASMA**.

The Great Pyramid is associated with the Taurus constellation and a powerful star known as Hyadum I, also called **PRIMA**.

5) The Plasma Reflux Pyramid – The Second Largest Pyramid on the Giza Plateau

The pyramid next door, the second largest on the Giza Plateau, is now more appropriately named the *Plasma Reflux Pyramid* after its function.

An overhead image reveals that the pyramid on the left has a limestone foundation significantly larger than that of the Great Pyramid on its right.

Given the design of this substantial foundation surrounding the immense structure, I determined that this system functions as a *heat sink*. It is designed to attract and absorb high-temperature energy, similar to the process used in electrical battery chargers.

When charging large battery banks, a great deal of heat is generated, which must be managed to prevent the system from overheating.

The Taurus constellation star associated with the Plasma Reflux Pyramid is **Theta Tauri**.

The Electric Pyramid is the newly designated name for the third and smallest pyramid on the Giza Plateau.

Out of all the pyramids, its name is the most distinctive.

From the base to halfway up its slopes, the Electric Pyramid is shrouded in red granite—similar to the material found inside the Great Pyramid's Reaction Chamber, which consists of 55% clear quartz crystal. Because of this, the Electric Pyramid could also be called the Piezoelectric Pyramid.

Piezoelectricity, remember, is **plasma**.

Among all the stars in the Taurus constellation, **Aldebaran** is the one that aligns with the Electric Pyramid. Aldebaran has a renowned ancient lineage spanning multiple civilisations. More recently, it has been known as **"The Follower"** because it always follows the Pleiades.

THE GIZA PARK PROTOCOL HOW TO START THE SUN

In Hinduism, it is called **Rohini** ("The Red One"). It is an **Alpha Tauri** star of significant luminosity, the fourteenth brightest star in the sky. Within the Taurus constellation, it is considered the **"Bull's Eye."**

The Electric Pyramid is a fitting name. It is the final structure to complete the architectural and energetic systems on the western border of the Giza Plateau.

Now, let us return to the Great Pyramid and examine the eastern structures and systems.

As we revisit the Great Pyramid, remember, **a bull has two horns.**

Above, you see an image of a female lioness. It is the initially carved representation of the Geyser Sphinx I drew.

I also like this representation someone else drew to reflect the original size of the Geyser Sphinx as a Lioness.

THE GIZA PARK PROTOCOL HOW TO START THE SUN

Following my in-depth research on this topic, I came across a text in England titled *The Jungle in Sunlight and Shadow*, written by British forester Frederick W. Champion.

After the war, Champion joined the Imperial Forestry Service in the United Provinces of India and gained fame in the 1920s as one of the first wildlife photographers and conservationists. As a wildlife researcher, he was the first to study and document in detail the diurnal hunting habits of female tigers, lions, and various other wild animals in their natural habitats, as well as their mating and gestation periods in India.

The female lioness symbolises the **295/70-day cycle** of the heliacal rising of Sirius on the **Summer Solstice**. This cycle aligns with the period when the lioness is **in heat**, through to the birth cycle, and the point when her cub at a maximum of three weeks old is introduced to the father.

Male lions are highly territorial and protective of their mates. If a mother delays introducing her cub to the father for too long, he may perceive the cub as a threat and kill it.

#7 Looking to our immediate left, we see the **Geyser Sphinx**. As shown in my images, our focus will be on the **18-inch hole** at the top of the Sphinx's head and its possible function.

THE GIZA PARK PROTOCOL HOW TO START THE SUN

THE GIZA PARK PROTOCOL HOW TO START THE SUN

THE GIZA PARK PROTOCOL HOW TO START THE SUN

The Geyser Sphinx is lucky number 7 for our identification and overview.

The Taurus constellation star associated with the Geyser Sphinx is known as SECUNDA, a HYDUM II strength star.

#8) Next is the Incredible Wall of the Crow, one of the most visible and massive mysteries on the Giza Plateau.

Soon, you will understand its design and function, like the rest of the systems.

THE GIZA PARK PROTOCOL HOW TO START THE SUN

You are looking at the top Lintel stone on the Wall of the Crow, the largest stone on the Giza Plateau.

Those 5x 70-ton red granite beams in the Great Pyramid inside the Reaction Chamber weigh less than the one lintel stone on top of the WOC.

THE GIZA PARK PROTOCOL HOW TO START THE SUN

The Taurus constellation star is associated with the Wall of the Crow, known initially as AIN, translated from Arabic, "The Bull's Eye."

This makes the second Eye of the Bull complete.

The more recent name EPSILON-TAURI is expected as well.

#9) Zawyet is another Military Base and off-limits for investigations. The images I offer are ones removed from the internet. You make your assumptions. It looks like a perfectly designed Chemical Mixing Containment system.

The Taurus constellation star most associated with Zawyet is TAU-TAURI.

#10) Another baffling underground structure is the Serapeum, a facility that contains massive-sized sarcophagus-like containers carved from stone in which legend says the Sacred Bulls worshipped were entombed.

THE GIZA PARK PROTOCOL HOW TO START THE SUN

THE GIZA PARK PROTOCOL HOW TO START THE SUN

THE GIZA PARK PROTOCOL HOW TO START THE SUN

THE GIZA PARK PROTOCOL HOW TO START THE SUN

THE GIZA PARK PROTOCOL HOW TO START THE SUN

THE GIZA PARK PROTOCOL HOW TO START THE SUN

The Taurus constellation star related to the Serapeum is **111-Tauri**.

#11) The Red Pyramid

The Red Pyramid was closed to tourists in 1967 due to strong chemical smells and reports of visitors falling ill after their visit. It remained closed for 30 years and was finally reopened in 1997 following extensive cleaning and ventilation.

During a nine-week, in-depth investigation in Egypt, Heidi and I worked exhaustively to ensure the accuracy and integrity of my theories. When we visited in 2022, I could still detect the scent of ammonia. Given the thousands of years of operations, my analysis suggests that liquid ammonia was produced over a span of **27,800 years**. That lingering smell may very well be embedded in the stone itself.

The Taurus constellation star most associated with the Red Pyramid is **Elnath (Beta-Tauri)**.

Elnath (Beta-Tauri) is a chemically peculiar star. It is a **mercury-manganese star** that is **non-magnetic**. This blue giant has a mass five times that of our Sun and is located at the tip of the Northern Horn of the Bull. It is the **second-brightest star** in the Taurus constellation.

I believe Elnath and its companion stars symbolise the **chemical production facilities at the Saqqara Chemical Facility**, which includes the **Step Pyramid, the Red Pyramid, and the Bent Pyramid**.

#12) Lake Moeris Fuel Supply System

See the images below—two artistic renderings depicting what is suspected to be the sinking of these two islands, with small pyramid structures still intact.

The Taurus constellation star most associated with Lake Moeris Supply Systems is TIANGUAN or ZETA-TAUR.

THE GIZA PARK PROTOCOL HOW TO START THE SUN

Crab Nebula

Supernova remnant

#13) Hawara Pyramid and Labyrinth

The Taurus constellation star does not apply here, as the **Crab Nebula** is directly associated with the **Hawara Pyramid and Labyrinth** rather than a specific star.

This concludes the identification of the stars and their mirrored image upon the ancient sands of the **Sahara Desert in Egypt**.

I structured these descriptions to clearly present the **Egypt–Taurus Constellation Correlation Theory**. In the next chapter, I

THE GIZA PARK PROTOCOL HOW TO START THE SUN

will begin with the **fuel source for the Great Pyramid**, providing a detailed explanation of the system's functions and how they operated in sequence.

Chapter Eleven:
Egypt Weather and Climate

Modification System
Part One
Includes Two Major Production Systems:

The Giza Industrial Park Complex comprises six primary systems:

Ancient Ionospheric Heater/Plasma Generator

Ground-Based Cloud Seeding Generator

Quantum Superconductive Plasma Electromagnetic Cyclotron

Ionic Plasma Reflux Heat Sink

Piezoelectric Plasma Excitor

Venturi Vacuum Generator

Chemical/Water Treatment Support Facilities consist of nine primary systems:

Lake Moeris, the source of seawater aquifers containing six mineral ions

A double-tunnel, manually operated supply system

A methane gas production system

Two ammonia production facilities—one producing liquid ammonia, the other powdered ammonia

A filtration and structuring facility

A chemical injection facility

The Giza Plateau Mixing Basin

THE GIZA PARK PROTOCOL HOW TO START THE SUN

Earlier, I discussed one of my most significant discoveries: that the Giza Industrial Park construction template mirrors the constellation Taurus, the Bull. However, it is essential to understand that I first had to identify each of these components before recognising its correlation with the Taurus constellation.

This is how the **Taurus Correlation Theory** was born.

THE GIZA PARK PROTOCOL HOW TO START THE SUN

See the star chart images below: m8- p/'

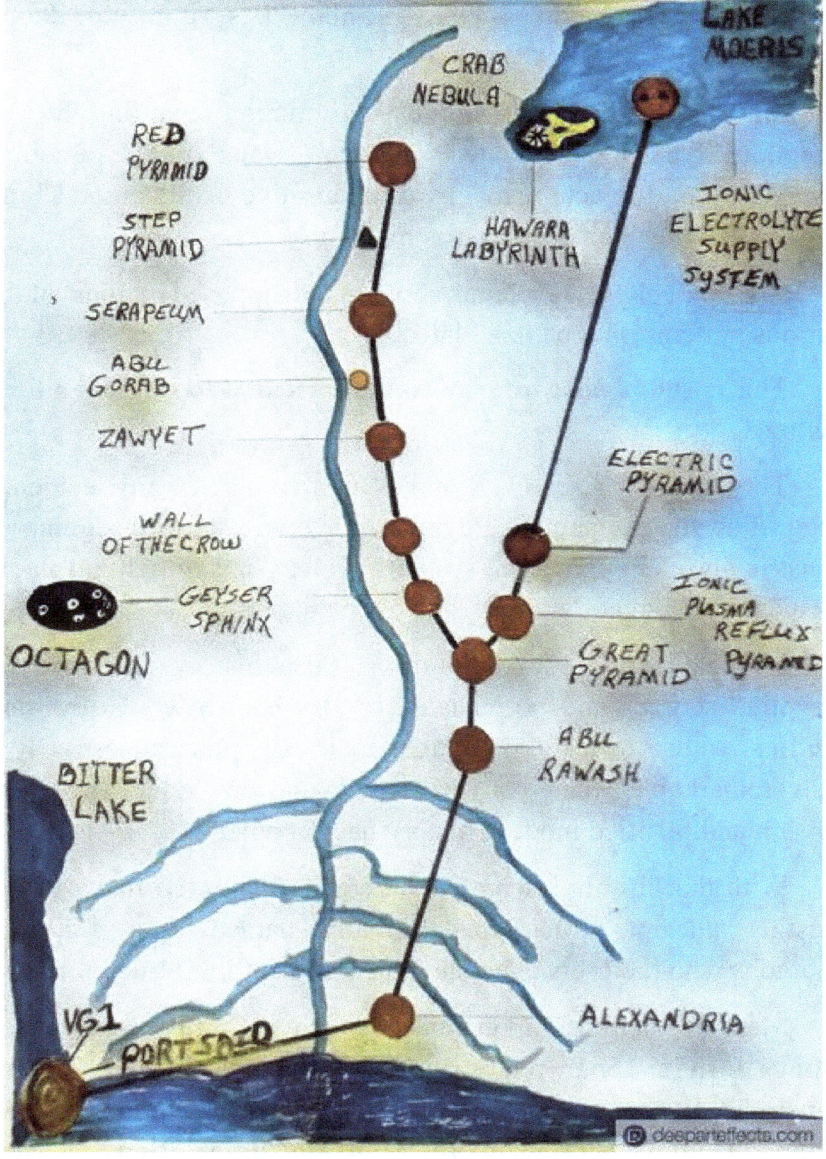

After reviewing the images, it is easy to see the precision of this **"As Above, So Below"** example—an embodiment of Hermes' philosophy in physical form. I repeat once more:

THE GIZA PARK PROTOCOL HOW TO START THE SUN

#3) Principle of Correspondence:

"As above, so below; as below, so above. As within, so without; as without, so within."

Since I have little trust in most information regarding the Old Kingdom dynasties associated with the Great Pyramid period in ancient Egypt, I reject the mainstream narrative that Pharaoh Khufu constructed the Great Pyramid as his tomb.

Nor do I believe the claims about other supposed builders of the various systems on the Giza Plateau.

This is **100% false information, fabricated to support a false history.**

Through my research, I have yet to uncover any evidence suggesting that an ancient Egyptian played a role in designing or constructing the operational systems of the **Giza Industrial Park**, including the Great Pyramid.

I repeat—**100% fabrication**—a false history promoted by Egyptian sources and perpetuated by **Jewish-owned Hollywood studios**, whose own religious texts claim Egyptian ancestry. It is truly remarkable when you realise, beyond any doubt, just how corrupt and falsified modern history has become.

With that in mind, I have deliberately **renamed** many of the existing ancient structures—originally named after Egyptian pharaohs who themselves had no knowledge of their true builders.

Many system components now have **industrial-inspired names**, while some retain their connections to the **Taurus constellation** or traditional deities and local references. This renaming allows for a more **accurate description** of each component's function within the overall system, distancing it from false narratives and providing a **realistic, functional framework** for understanding these systems.

THE GIZA PARK PROTOCOL HOW TO START THE SUN

For example, to name just a few:

The **Great Pyramid**—previously misattributed to Khufu (Cheops) will continue to be called the **Great Pyramid**, as it remains the only surviving structure among the original **Seven Wonders of the World**.

Its revised nomenclature will directly identify it with the **Taurus constellation**.

Old Names

Giza Pyramids, Pharoah's Tombs, Kufu, Cheops, The

Great Pyramid.

New Names

Egypt Weather & Climate Modification System/Giza Industrial Park

The Great Pyramid remains the designated ID.

Taurus Constellation ID, Star – Gamma-Turi/First Hyad-also Prima.

Old Name

Kafrae

New Name

Ionic Plasma Reflux Pyramid.

Taurus Constellation ID, Star- Delta Tauri/Second Hyad

Old Name

Menkaure

New Name

Electric Pyramid.

THE GIZA PARK PROTOCOL HOW TO START THE SUN

Taurus Constellation ID, Star-Aldebaran/The Torch Bearer/The Follower/The Fiery Eye of the Bull

Old Name

Wall of the Crow

New Name

Venturi Vacuum Generator

Taurus Constellation ID, Star-Epsilon Taui/Second Eye of the Bull

Old Name

Sphinx

New Name

Geyser Sphinx

Taurus Constellation ID, Star-Theta Tauri

Old Name

Old Kingdom Egypt Temples & Pyramids

New Name

Chemical and Water Treatment Facilities

Old Name

King's Chamber

New Name

Reaction Chamber

Old Name

Pharoah's Sarcophagus

New Name

Piezoelectric Direct Discharge/ Plasma Transformer

THE GIZA PARK PROTOCOL HOW TO START THE SUN

Old Name

Osiris Tomb   unchanged

New Name

Osiris Tomb, however, has a system designated ID

Quantum Superconductive Plasma Electromagnetic Cyclotron

Old Name

Queen's Chamber

New Name

Now that I have detailed many systems and briefly explained their functions, I want to explain how I feel about the original builders of these massive structures.

I believe there existed super-intelligent ancient Master Builders of the Universe—extreme geniuses who understood the most intricate details of our planet, earthlings, and the entire cosmos.

Past civilizations, in this case, the Ancient Egyptians, were among many primitive, wandering tribes in the desert who relocated to this region after the ice melted.

These primitive people did not have the knowledge, resources, finances, or methods to suddenly appear in the perfect location in the desert, then design and construct an incredible system to increase the incidental radiant heat temperature of the Earth.

"Terminating the Ice Age."

It involves creating and building a system composed of all natural systems to provide a suitable environment for humans.

Multiple natural systems were combined to influence the weather and climate of the Earth. We have discovered that at least five ancient pyramid systems operated together in unison. Each of these systems had a distinct function and was intentionally designed and installed to generate heat and provide the necessary water. They could have terminated the Ice Age in a shorter, more productive, and controlled manner.

As we embark on this journey together, I am 100% convinced—based on my discoveries alone—that the three primary functions of the Giza Industrial Park are: the Giza Industrial Park Support Systems, the infrastructure necessary to provide seawater, water filtration, water structuring, disassociation, and chemical additives.

These procedures are essential to adequately prepare the water before it enters the initial processes of the Giza Industrial Park system, specifically the multi-component system of the Great Pyramid.

Below is a flowchart illustrating the 20 dedicated primary systems of the weather and climate modification components of the Giza Industrial Park, including the Water Treatment Support Systems, with a brief description.

The ancient Egyptian weather and climate modification systems flowchart was designed to "control" the termination of the Ice Age, ensuring human survival and prosperity. Evidence confirms that humans migrated and inhabited every continent except Antarctica after the Ice Age, which lasted between 2.4 to 3 million years.

This brilliantly designed system must have been constructed much earlier than the 14,500 BC date proposed by scientists for the melting of the Ice Age. The pyramid systems would have needed to be activated long before that significant event—the Bonneville Flood.

My research into dating these events is based on an extensive compilation of scientific research and testing. Fortunately, I have analysed vast amounts of data and have decades of experience in forensic investigation, coupled with expertise in industrial mechanical support systems, high-pressure turbine pump rebuild engineering, and marine nuclear piping systems analysis.

This expertise allows me the freedom to confidently apply reverse engineering methods to understand and recreate ancient systems, as I have done with this treatise. Modern scientists use the latest research from ice core samples collected from the northern and southern regions of the Earth.

Scientific data indicates that around 14,500 BC, the 2-mile-thick ice sheet in North America began to melt. Further studies suggest that by 13,700 BC, the average sea level had risen by approximately 30 feet.

By harnessing the power of this indigenous, natural, self-sustaining, eco-friendly solar-powered system, the creators possessed knowledge of cosmic energy systems that we have yet to fully comprehend.

My goal is to present a fresh perspective on how the universe operates and how these natural cycles impact us as a species and our environment.

Today, much of modern science is only as advanced as the latest technology available to measure and test resonance, frequencies, and vibrations.

Consider HAARP systems, CERN particle accelerators, directed energy weapons (DEW), and weather and climate modification technologies. We are only now beginning to understand many of Nikola Tesla's works that were not lost or stolen after his death. He correctly predicted that his inventions would make him world-renowned within 100 years, and he has been dead for over 80 years.

Furthermore, scientific integrity must be preserved.

For instance, on 19 September 2019, the sitting president of the United States enacted Executive Order 13887, which removed all existing safety protocols for vaccine research. This effectively granted pharmaceutical companies unprecedented protection against liability for vaccine-related injuries or deaths. As a result, it became illegal for affected individuals to even pursue legal action in a US courtroom.

I am pleased for them—they have sold thirteen and a half billion experimental vaccines. However, the pharmaceutical industry has fallen under the control of profiteers, making it one of the most corrupt sectors. Publicly released research results are often dictated by funding sources and their vested interests.

Conversely, as a true scientist, I uphold dynamic and evolving research. My hypotheses and theories shift as new evidence emerges. I do not base my discoveries on a predetermined conclusion or financial incentives.

My aim is to provide sufficient new information from various perspectives, allowing you to assess the evidence and form your own conclusions. By doing so, you will be able to recognise and interpret my presentation, engage with it, and perhaps even participate in further research.

Understanding ourselves and our environment requires the development of new techniques and infrastructures.

I have identified and systematically arranged a procedural set of instructions, complete with a flowchart and technical drawings of all the weather and climate modification systems. These illustrations provide a broad overview, including a visual layout of the extensive underground tunnel systems. The diagrams are systematically structured to encompass all above-ground systems, pyramids, and supporting infrastructure.

I have identified 20 dedicated systems that comprise the entire Egyptian weather and climate modification system.

The Water Treatment Support System is further divided into two specialised subdivisions. Together, they establish the supporting infrastructure necessary to provide and prepare ionised seawater for use as fuel. This specially filtered, structured, chemically treated, and disassociated seawater is then utilised in the Chemical Injection System, where these two independent systems converge, allowing alchemical processes to take place.

Prior to this, the treated water is transported to the Mixing Basin at the Giza Industrial Park Complex. Here, several additional processes enhance the water's structure, including resonance, sonic, electromagnetic field (EMF) exposure, piezoelectric effects, and plasma treatments.

The Giza Plateau's life-giving water supply provides the essential foundation for the three major operating systems.

THE GIZA PARK PROTOCOL HOW TO START THE SUN

Very rare, pre-1851 engraving of the two pyramids with colossal figures atop each. Accounted by Herodotus as being located in the center of Lake Moeris, Egypt 490 BCE.

THE GIZA PARK PROTOCOL HOW TO START THE SUN

I aim to offer something that has not been done in modern history. I am providing an identification and flowchart of all operational systems of the Ancient Egyptian Weather and Climate Modification System.

My second objective is to include a step-by-step procedure for systems analysis and operating instructions.

The manual operating instructions on "How to Start the Sun" are provided in the volume you are reading.

Flowchart

This flowchart begins at the source—an artificially excavated lake designed as part of this ancient industrial complex. Lake Moeris provides raw seawater from underground, mineral-rich aquifers with a high negative-ion content. This unique water is captured and forwarded to two water treatment systems for further processing.

#1) Lake Moeris - Water Distribution System

THE GIZA PARK PROTOCOL HOW TO START THE SUN

Lake Moeris serves as the source or supplier of raw seawater from deep underground aquifers.

Two small pyramid-like structures were constructed near each other, most likely on a small island near the shoreline but surrounded by water. These two small pyramids each have a statue on top.

An old sketch illustrates what I am about to describe. First-hand satellite archaeology has confirmed that these two obelisk-like structures are still located in the same place but now lie very deep beneath the water.

This is a vital clue in explaining the ancient stories from various civilizations about a great flood that destroyed much of life in the affected areas. Returning to the two obelisk structures—while they are identical and perform the same function, they are independent systems.

Each obelisk can open and close an intake valve, allowing Lake Moeris seawater to rapidly flood into two tunnels. These obelisk gate valves can adjust the flow incrementally based on system requirements, providing excellent operational flexibility.

The obelisks also play a crucial role in shutting off the Weather and Climate System for its annual maintenance. The valves are reopened before the heliacal rise of Sirius to initiate the startup and operational cycle, which lasts for 295 days a year. The two water distribution obelisks are also designed to quickly halt water transport into the tunnels in case of an emergency.

Raw seawater serves as a transport medium containing six mineral ions, which act as fuel for the processes required to alchemically transform the water. These processes include mineral structuring, chemical injection, resonance frequency vibrations, electromagnetic and EMF cyclotronic stimulation, impact forces, sonic frequencies—including piezoelectric effects—and water dissociation.

THE GIZA PARK PROTOCOL HOW TO START THE SUN

All these processes are necessary in the initial stages of the system. Multiple methods are employed to separate the water's hydrogen and oxygen molecules, ultimately producing plasma in the final phase.

Following this extensive treatment, the water passes through approximately 50 miles of limestone bedrock tunnels. The dynamic movement generates kinetic energy and extracts carbonate minerals from the limestone into the liquid.

As the treated liquid moves, it gains kinetic energy and increases its scalar conductivity potential. While all these processes are crucial in creating the required chemical environment, their primary purpose is to dissociate the water—separating its oxygen and hydrogen molecules to generate hydrogen gas, which is essential for producing plasma.

Seawater generates electric charges and closely resembles human blood plasma chemistry.

Gravity-fed tunnels control the movement of this liquid, directing it to two water treatment facilities.

As mentioned earlier, I will briefly introduce the 20 different components of the Egyptian Weather and Climate Modification System. As we navigate this ancient maze of ingenuity, I will elaborate on key elements as necessary.

I will begin with:

#2) Pyramid Operations Water Control (POWC)

This system includes chemical facilities responsible for methane gas production, liquid ammonia production, and powdered ammonia production for future chemical injection alongside #1) (POWC).

Note: The production of methane gas and ammonia at the chemical manufacturing systems follows a different schedule, beginning before the summer solstice. Additionally, powdered fertilizer production at the Dozer Pyramid operates continuously with routine maintenance schedules, independent of the Weather and Climate Modification System's operational periods.

Once the command is given to operate the Pyramid Control Valve, seawater is allowed to flow by gravity into Tunnel #2. This tunnel stretches for miles before reaching its destination at the Step Pyramid of Saqqara. As the water moves through the tunnel, it becomes highly agitated, increasingly dissociated, more electrically charged, and primed for the next phase of processing.

#3) Step Pyramid of Saqqara—Methane Gas Production Facility

The facility produces methane gas through the fermentation of agricultural waste and cattle manure. Bacterial digestion generates methane as a byproduct of decomposing cow manure. This process involves several steps, requiring the facility to occupy 40 acres of land to meet demand.

Gravity-fed tunnels then transport the methane to the Red Pyramid of Dahshur.

#4) Red Pyramid of Dahshur

This pyramid system was closed to the public in 1967 after numerous visitors complained about the strong ammonia smell and became ill. It remained closed for three decades before reopening in 1997 following an extensive cleanup.

I agree with Geoffrey Drums in his text *The Land of Chem* that the Red Pyramid functioned as a chemical industrial facility. It utilised the methane gas supplied by the Step Pyramid to generate heat, producing a chemical liquid called ammonia.

Once this multi-stage process is complete, the Red Pyramid has two gravity-fed exit tunnels. One tunnel flows southwest, transporting the liquid ammonia produced by the Red Pyramid into the **#5) Bent Pyramid**. The second tunnel runs north and south, distributing liquid ammonia to the Bent Pyramid, where it is chemically converted into powdered ammonia for fertiliser. This fertiliser was used to support crop growth and grain production, which in turn sustained the cattle needed for manure production.

For this reason alone, I believe that, over time, the cow became the sacred bull—specifically, the Apis Bull, black with a white diamond on its forehead. The Apis Bull was the most royal and symbolically represented the Pharaoh within the animal kingdom.

I once had the opportunity to pet an Apis Bull in India. It would not surprise me if genetic analysis revealed that the Apis Bull was brought from India to Egypt, much like the Blue Lotus. While Egypt claims the Blue Lotus, it is not indigenous to the region—the Yellow Lotus is. Perhaps the Apis Bull and the Blue Lotus arrived in Egypt together during the same period?

Neither cows nor camels are indigenous to the Sahara Desert regions of Egypt; both species were imported.

Additionally, an Egyptian policeman riding a camel once shared an insight with Heidi and me regarding the Egyptian perspective on camels. He explained that he had been riding the same camel for three years, and the animal's service time as a police mount would expire the following year. When Heidi asked what would happen to the camel afterward, we were both stunned by his response.

Smiling as he affectionately patted the camel's neck, he told us he would **take it home and eat it**.

It is clear that in Egypt, there are only sacred cows—if they are all sacred kebabs.

Conversely, in India, the Apis Bull and all other cows are revered. They are never slaughtered for food; instead, their milk is used to produce ghee, which plays a crucial role in spiritual rituals. One such ritual involves casting ghee into an open fire. The widespread practice of this ritual across different cultures demonstrates its global significance.

THE GIZA PARK PROTOCOL HOW TO START THE SUN

THE GIZA PARK PROTOCOL HOW TO START THE SUN

The other north tunnel carries the finished product—liquid ammonia—directly to **#7) Zawyet El Aryan**, a chemical mixing facility.

The mixing basin requires a constant and carefully controlled fuel level, including a control centre that regulates the fuel supply being forwarded north to the **Giza Industrial Park**. There are several reasons for this.

The **Cadman Pump** is located 100 feet below the ground of the **Great Pyramid**. My friend John Cadman, also known as the "Mashed Potato Guy" for his work on what was formerly called the Pharaoh's Pump, has conducted exemplary research on this subject. I take his findings as fact, not only due to his expertise but also because of my own experience. As a **Vertical Pump Project Engineer**, I was responsible for removing and reinstalling the **fire suppression pump for the World Trade Center** before 9/11. This system played a crucial role in saving many firefighters and allowed upwards of 50,000 people to escape from the Twin Towers—both 110 stories tall—on that tragic day.

For this reason, I highly recommend that everyone visit **John Cadman** and witness his **functional perpetual pump system**, which replicates the ancient Egyptian mechanism on a smaller scale. I know firsthand that John's pump operated for approximately 15 years **without electricity**. It is a **100% self-perpetual system**.

You will understand this better when we reach the **Ignition Start-Up Phase**.

Maintaining the correct **water (fuel) level** was critical. If the level was too high, the **vacuum-type intake system** of the **Cadman Pump** could become flooded, causing it to miss strokes. This, in turn, could create **hydraulic water rams** inside the **Great Pyramid**, ultimately disrupting its ability to generate **plasma**.

THE GIZA PARK PROTOCOL HOW TO START THE SUN

Conversely, if the **water (fuel) level** was too low, the pump would experience **fuel starvation**, impairing its function.

Both of these scenarios could result in the **Cadman Pump shutting down**, and it could only be restarted during the next **Summer Solstice**, coinciding with the **heliacal rise of Sirius**.

This is yet another example of why the **Summer Solstice** was not only crucial for Egyptians living in the **Sahara Desert** but also for **humanity as a whole**, as it played a role in **terminating the Ice Age that lasted 2.4 million years**.

#5) Bent Pyramid of Dahshur

While the **Bent Pyramid** also functioned as a **chemical industrial facility** that utilised the **liquid ammonia** produced by the **Red Pyramid**, its role was not directly connected to the agenda of **terminating the Ice Age**.

Instead, its primary function was to **convert liquid ammonia into a powdered form**, which was then used as **fertiliser for crops**.

As mentioned earlier, the **Bent Pyramid** was not directly involved in this larger process, but it played an **indirect role**. Agricultural food production was essential to sustain both the **labourers** and the **cattle**, whose **dung** was required to generate the **methane gas** necessary for ammonia production.

THE GIZA PARK PROTOCOL HOW TO START THE SUN

Situated on the **banks of the River Nile**, the **Bent Pyramid's** location suggests that **boats** were likely used to **transport powdered ammonia-based fertilisers** throughout **Egypt**.

With this, we now conclude the **operational design and responsibility** of the **#2) Pyramid Water Control System**, as it relates to the **chemical production process**.

Let us now return to **Pyramid Water Control #1)**, which I initially skipped. You will soon understand why.

#1) Pyramid-Water Operations Control – Structured/Seawater Plasma System

This is the **primary water supply and transport system** that powered the **entire Giza Industrial Park**.

When the **valve open command** was activated, raw **seawater**—just as described in **System #2**—was directed **downward into Tunnel #1**. Gravity then forced this liquid to flow toward its **first destination**…

THE GIZA PARK PROTOCOL HOW TO START THE SUN

#6) Serapeum of Saqqara is the first-stage water treatment facility that provides filtration, mineral, and resonance frequency water structuring. The water is then gravity-fed directly to #7) Zawyet El Aryan for mixing with the Liquid Ammonia from the Red Pyramid.

THE GIZA PARK PROTOCOL HOW TO START THE SUN

Part Two

BOSNIAN & EGYPTIAN PYRAMIDS CORRELATION THEORY AND START-UP PROTOCOL FOR THE EGYPT WEATHER & CLIMATE MODIFICATION SYSTEMS

Once the Chemistry is completed at Zawyet, prepare the desired structure and frequency of the intake water for the Great Pyramid to begin its magic. These systems will soon become operational with the Bosnian Pyramids Systems, operating in Tandem.

Fouilles de Zaouiét-el-Aryân. — Déplacement des blocs de granit
(Février 1912).

#7) Zawyet El Aryan is a chemical injection facility where liquid ammonia is produced at the #4) Red Pyramid and dispersed incrementally into the gravity-fed, structured seawater flow stream provided by the #1) Pyramid Water Treatment Control.

THE GIZA PARK PROTOCOL HOW TO START THE SUN

The chemical injection process is controlled and measured precisely. Once the liquid ammonia has been properly dispersed, the concoction is metered and gravity-fed into tunnels that drain into the Mixing Basin at the Giza Industrial Park, in front of the Geyser Sphinx System.

#7) Giza Industrial Park Mixing Basin is a large, excavated retaining pond for the collection and storage of an adequate supply of treated seawater from Lake Moeris. This includes structuring, the injection of measured liquid ammonia, and plasma waters from the Ionic Plasma Reflux Pyramid's exit causeway #2, in preparation for being supplied through the intake tunnel of the Cadman Pumping System, which is installed over 100 feet directly beneath the Great Pyramid.

The Mixing Basin also receives structured and treated water from the Plasma Reflux Pyramid causeway overflow plasma water, as well as the Geyser Sphinx overflows through the Geyser Sphinx Water Structuring Chamber.

The Great Pyramid's Cadman Pump is the **heartbeat**, the **pulse** of the Giza Plateau and beyond. As the pump intakes the treated seawater, suction creates a vacuum responsible for moving all the necessary seawater throughout the Giza Industrial Park systems. The Great Pyramid and all other systems at the Giza Industrial Park function both beneath and above the ground.

Engineers calculated this pump's performance capabilities with a design limitation of 3,200 psi head pressures. The pump operates at 60 strokes per minute—one stroke every second.

The entire Giza Plateau would rock and shake with vibrations from the force created by this enormous hydraulic perpetual pump. Additionally, the Cadman Pump generates a strong resonance frequency vibration that travels through all the systems and extends for miles through the solid limestone bedrock beneath the entire Weather and Climate Modification System.

THE GIZA PARK PROTOCOL HOW TO START THE SUN

All maintenance has been completed over the past 70-day shutdown cycle. The systems have now been adequately prepared and primed. The work crews are on-site and safely prepared to begin the start-up of the Giza Industrial Park.

The appropriate amount of processed fuel has been prepared, and the Mixing/Collection Basin has reached its maximum capacity, sufficient to begin the first stage.

It is time to begin.

The two vessels and crews must manually lift the massive granite block—the **on and off switch** for the start-up of the Great Pyramid, the workhorse that creates the required magic. Giza Park, afloat in the Mixing Basin, serves as a safety valve and relies on a well-organised and carefully executed plan.

The Solar Boat below is one of the vessels discovered stored in one of the six boat pits at the Giza Plateau.

I believe these vessels were stored there as part of the equipment needed to operate the Giza Industrial Park and the Egypt Weather and Climate Modification Systems when not in use.

THE GIZA PARK PROTOCOL HOW TO START THE SUN

THE GIZA PARK PROTOCOL HOW TO START THE SUN

Below is one of the empty boat pits by the Great Pyramid.

I'm using this image to show the vessel's attitude when it is secured and lifting the heavy granite block used to cover the tunnel intake for the Cadman Water Pump installed 100 feet beneath the Great Pyramid.

Now that you know how the Cadman Pump is activated, I can explain the events more clearly.

THE GIZA PARK PROTOCOL HOW TO START THE SUN

#9) The Great Pyramid is the only remaining structure from the original Seven Wonders of the Earth. It is the Great Pyramid for many reasons, more than I will detail now: its sheer size, its remote location, and its perfect geological and energetic alignment at the precise centre of the Earth.

Its unique design possesses capabilities that no other pyramid system can compare with. The materials used in its construction were chosen for specific purposes. It utilises indigenous materials to capture over 100 quantum energy streams, combining them into multiple systems to modify weather and climate globally and to end the Ice Age.

Paired with the Plasma Reflux Pyramid, these two systems were significant components of the Ionospheric Heater Plasma systems. Now would be the perfect opportunity to discuss precisely what plasma and scalar energies are.

When considering its intended function, one of the most remarkable discoveries about the Great Pyramid led me to believe that one of its abilities was the production of hydrogen gas.

At first, I was convinced that the primary role of the Great Pyramid was to generate hydrogen gas. I had accumulated a substantial amount of collaborative evidence to support this theory. However, over time, I came to realise that I was both correct and incorrect at the same time.

I have since compiled a detailed definition of plasma, along with some natural and artificial examples.

First, I will quote my friend from New Zealand, Veda Austin, who is doing extraordinary work by freezing water and capturing physical images of "consciousness" frozen in ice. Veda has amassed a vast collection of water manifestations. Additionally, she has gathered several profound quotes and definitions regarding water.

Veda claims that water is the largest liquid source on Earth. When we consider water's abundance, we assume it is indigenous to Earth and has always existed here. That assumption, however, is entirely incorrect.

When discussing Earth's water, we must once again look into the Cosmic Mirror, as I call it. The water on Earth is vast, but it arrived from the stars, stored within meteorites and asteroids that impacted the planet multiple times over 4.5 billion years ago.

Water is not merely a resource; it is the source and responds intelligently. Water contains living messages from the universal consciousness.

Liquid is the body of water.

Hydrogen is the spirit.

Plasma is the brain.

Seawater chemistry is strikingly similar to human blood plasma. Like an electrolyte, it has the ability to generate an electrical charge.

Plasma is produced when matter is superheated—so hot that electrons are stripped away from atoms, forming an ionised gas. Plasma makes up over 99% of the visible universe. In the night sky, plasma appears in the form of stars, nebulas, and even the auroras that sometimes ripple above the north and south poles.

I love Veda's comment: "Water is the glove on the hand of consciousness."

Physics, on the other hand, describes plasma as an electrically conducting medium containing roughly equal numbers of positively and negatively charged particles. It is created when a gas becomes ionised and is sometimes referred to as the fourth state of matter—distinct from the solid, liquid, and gaseous states.

Images:

See the four elements below: solid, liquid, gas, and plasma.

As I stated before, hydrogen is essential for creating plasma.

The Greeks defined plasma as a "mouldable substance." It is the fourth phase of the fundamental states of matter: solid, liquid, and gas. A portion of that gas medium contains a significant volume of charged particles.

These charged particles form plasma and can consist of any combination of ions or electrons.

Plasma is the most abundant form of ordinary matter in the universe. The highest concentrations of plasma exist primarily in stars, including our Sun.

THE GIZA PARK PROTOCOL HOW TO START THE SUN

A standard method for creating plasma involves filling a glass bulb with gas and passing AC electricity through it. We are all familiar with these devices.

Below is a diagram illustrating that human blood consists of 55% yellow plasma. Additionally, I want to emphasise that male sperm is composed of 95% plasma, with the remaining 5% consisting of testicular hormones, which determine the sex of the fertilised egg.

This fact alone serves as 100% physical evidence of plasma's significance in the creation of life forms and abundance.

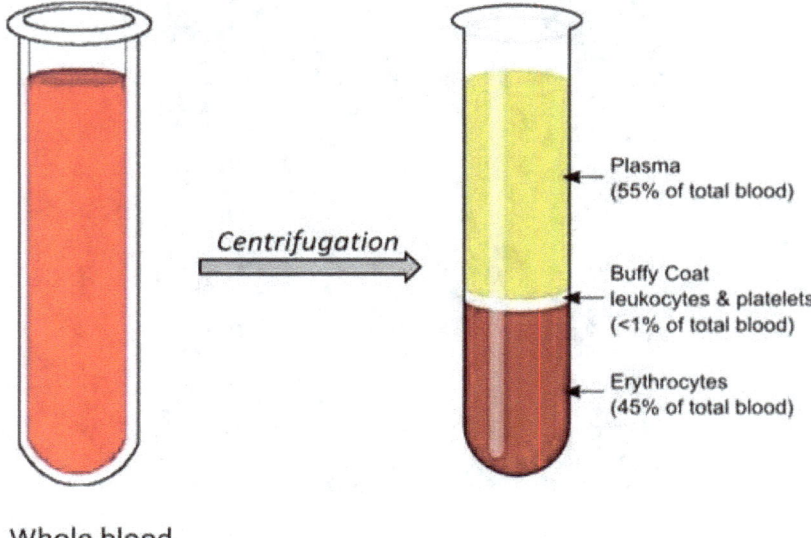

Whole blood

I will repeat Veda Austin's definition: **"Plasma is the brain of hydrogen gas."**

As I stated above, my earlier conclusions regarding the Great Pyramid's production of hydrogen proved to be both correct and incorrect.

Ultimately, I determined that the Great Pyramid's primary function was to generate plasma. However, to create plasma, hydrogen gas must first be produced. Once I discovered this missing equation, I immediately revised my theories. As an honest scientist dedicated to truth, I believe that when new scientific research provides fresh insights, theories must evolve to accommodate the latest data.

Plasma can be artificially generated using various methods, including heating, creating a strong electromagnetic field, or electrolysis, which involves high-voltage electricity or a piezoelectric current.

THE GIZA PARK PROTOCOL HOW TO START THE SUN

Lightning strikes are a form of plasma, as is the piezoelectric discharge from clear quartz crystals when mechanically stimulated.

The frequency range of lightning bolts, piezoelectric discharges, and the Earth's ionosphere all fall within the same very low-frequency (VLF) range, fluctuating between 3 kHz and 30 kHz.

Since piezoelectricity is a significant component of Egypt's weather and climate modification system, I want to explain it in a way that is easy to understand.

Below is an example of a typical gas lighter, similar to the one used for lighting a barbecue grill, candles, or a fireplace.

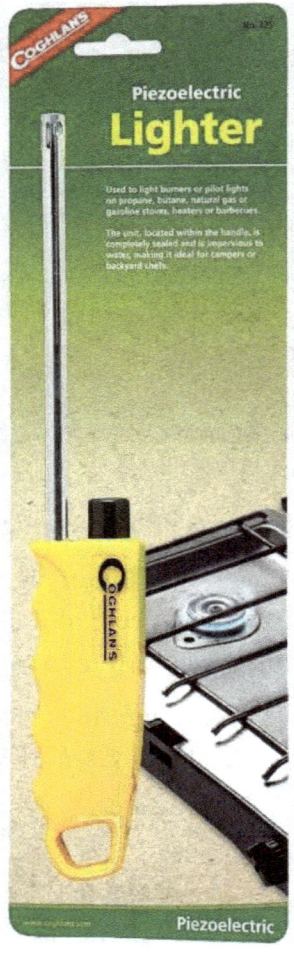

Most folks believe these lighters create a spark of electricity from a battery to ignite the gas. However, they need to be corrected.

THE GIZA PARK PROTOCOL HOW TO START THE SUN

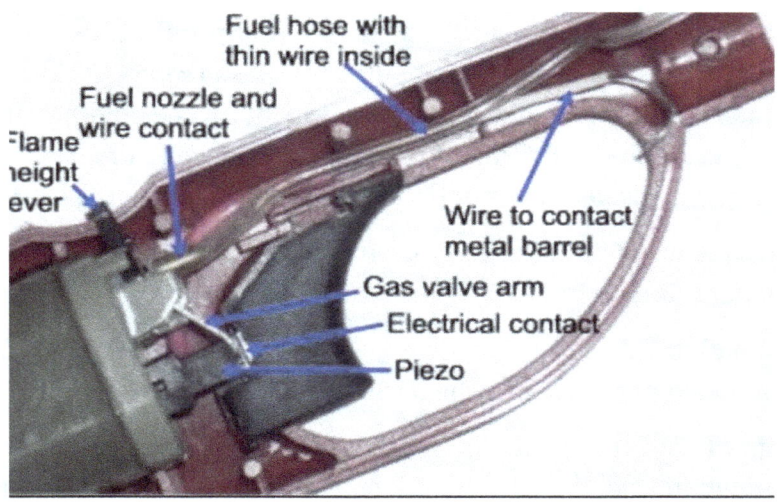

The diagram above shows the different components that make up the operating mechanisms of the typical grill gas lighter.

Below is an image I drew 15 years ago comparing it to the Electric Pyramid, which is appropriate.

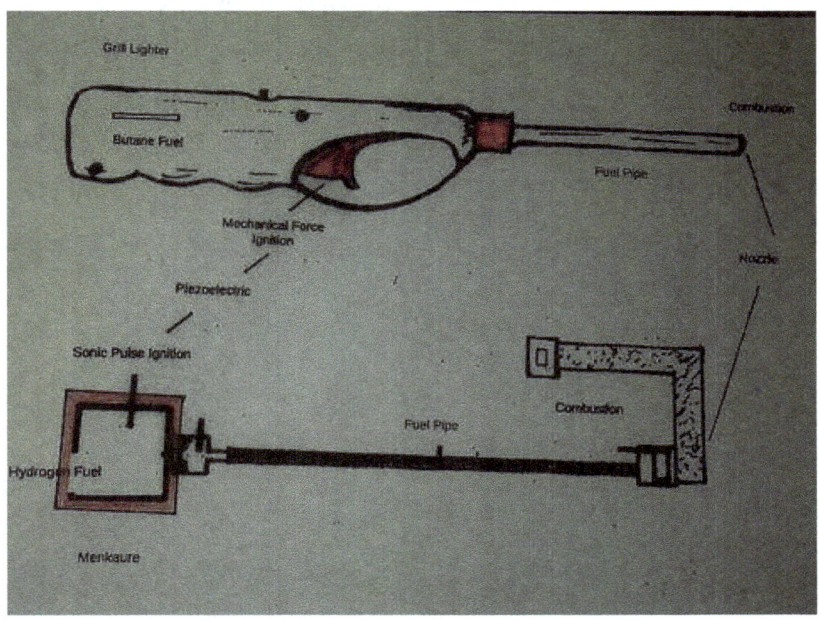

THE GIZA PARK PROTOCOL HOW TO START THE SUN

As I write this update, I am following up on my two recent forensic investigations, which involved weeks of visits to Bosnia. I examined physical evidence, reviewed many of the findings from 11,000 scientific experiments, and experienced firsthand the overwhelming scientific proof at the Bosnian Pyramids.

Dr Semir Osmanagić has made the greatest discovery on Earth in modern times. He deserves the Nobel Prize in Science.

I call any scientist who denies my claims a liar and a thief, as they are denying humanity the truth.

My final determination was made only after a thorough forensic examination and the identification of physical and collaborative evidence. This required two visits and the opportunity to work in the field alongside other scientists who conducted specialised scientific measurements. Their instruments were designed to gather precise data, which can be easily verified within an honest scientific system.

This is an ideal moment to detail two events I personally experienced, which are particularly relevant to this discovery.

For over a decade, I have conducted countless tests on ancient EMF waveguide systems from India, Buddhism, and Egypt. Dorji's Vajras evolved from sceptres into rods.

THE GIZA PARK PROTOCOL HOW TO START THE SUN

I like this collage I made after taking a photo of a dandelion.

In 2020, I began travelling the globe with these devices, measuring energy fields. By 2023, I had taken my research a step further, incorporating an ambulatory EEG device to measure the influence of brain waves at ancient sites, energy vortexes, temples, and pyramids.

THE GIZA PARK PROTOCOL HOW TO START THE SUN

During the in-field conference, I climbed the Pyramid of the Sun alongside other scientists to collect samples, as recently shared by Dr Semir. I brought my waveguides to conduct my own tests, and this is my report.

On the morning of **6th September 2024**, Heidi and I met with the other scientists and participants before travelling to the base of the Pyramid of the Sun. Each of us had our own agenda, carefully planning and executing our ascent up the treacherous, steep, and slippery slopes of this massive pyramid.

Upon reaching a **4-square-metre** area designated for sample collection, we paused to rest, orienting ourselves to the new environment and reflecting on the effort it had taken to get there.

THE GIZA PARK PROTOCOL HOW TO START THE SUN

The blue circle above represents the 4-metre area determined using various measuring devices and instruments, from pendulums to the most advanced electronic equipment available today. This is how many scientists have identified the exact location where this energy is being transmitted from the top of the Bosnian Pyramid of the Sun.

That day, I allowed each scientist to take their desired measurements before I began my testing. Eventually, I gathered my devices and stepped into the centre of this 4-metre area, said to be the top of the Pyramid of the Sun.

I typically orient myself to face east, a practice I have developed over the years. This orientation tool allows me to measure existing fields and detect the most common sources of interference. I refer to tall buildings, walls, water, and open fields—each environment has a distinct feel.

Almost immediately, I sensed something unusual. I usually receive energy streams parallel to the Earth's surface, meaning they are horizontal. Initially, I feel sensitivities in my hands and arms, which then spread throughout my body if a strong field is present. Typically, I locate the north and south polarity streams, as they are the strongest and serve as a reference for calibration. However, on this occasion, I could not detect any horizontal EMF streams, which was highly unusual.

This suggested to me that the vertical energy was more potent than the EMF field interference.

The energy felt rough and uncomfortable, shooting up vertically from the ground. I could feel it rising through my feet and into my legs. My body quickly reacted, becoming stiff and rigid from my lower extremities upwards.

Within minutes, I abandoned the experiment.

THE GIZA PARK PROTOCOL HOW TO START THE SUN

My body felt as rigid as steel, and I became dizzy, disoriented, and confused. I felt unwell and did not eat for the rest of the day.

In the days that followed, the sensations I had experienced seemed oddly familiar, and I believe I discovered why.

It was the summer of 1996, and I was living on board my 38-foot sailboat, *BURGOO*. I was wearing my bathing suit and had walked up to the boatyard to speak with some fellow boaters. In the distance, we could hear thunder, signalling an impending storm.

Suddenly, a torrential downpour began. Everyone ran for cover—except me. It was the first warm rain of the season, and it felt magnificent. I decided to return to my vessel and enjoy a natural shower, something I was accustomed to doing.

Once on board, lathered up and enjoying the moment, I struggled to rinse the soap from under my arms due to the wind direction. To solve this, I moved forward of the mast, raised my arms to the sky, and relished the experience.

Then, everything happened at once—fast, intense, deafening.

My body became as stiff and rigid as steel, my chin forced down into my chest.

Still standing, I watched as the electronic wind meter from the top of the mast landed beside me, and the wire cables caught fire.

Though my balance remained stable, my mind was thrown into a state of delirium that lasted for hours. I wandered aimlessly, dazed and disoriented.

That evening, my body was consumed by aches, and I eventually fell into a restless sleep.

By the next day, however, I felt invigorated. As a runner, I took to the road and performed as if I had just won a race. At the gym, I increased my weights effortlessly. And then—just like that—it was gone.

THE GIZA PARK PROTOCOL HOW TO START THE SUN

Minutes ago, I emailed Dr Semir, who is currently in India, requesting the readings from the different samples the scientists had observed at the top of the Pyramid of the Sun.

His reply was as follows:

"The electrical, magnetic, and ultrasound frequency at the top of the Pyramid of the Sun was measured at 28 kHz."

Three different scientists—one measuring electrical fields, another measuring magnetic fields, and the third measuring ultrasound frequencies—all reached the same conclusion.

This 4-metre squared area, which coincidentally matches the exact 4-metre measurement of the top of the Great Pyramid, registered a frequency of 28 kHz. This is the same range as lightning bolts, piezoelectric discharges, plasma, and the Earth's ionosphere's electromagnetic radiation in the VLF range of 3 kHz to 30 kHz.

My body felt horrible after testing the energy stream from the top of the Pyramid of the Sun because it remembered what lightning felt like.

THE GIZA PARK PROTOCOL HOW TO START THE SUN

Another incident I encountered not long ago involved using my new small-capacity ultrasonic jewellery cleaning device. I had never used one before, but it came with a pair of tweezers to allow me to pick the jewellery out of the hot chemical cleaning tank. The device claimed it could clean jewellery in three minutes, so I wanted to see the results.

For context, in 2016, I required a total shoulder replacement of my right shoulder. The metal components installed during surgery were fabricated from the highest quality titanium steel.

Being unfamiliar with the new ultrasonic cleaner, I reached in with the tweezers to inspect a silver bracelet while the device was plugged in and operating. I got a sudden jolt of energy up my arm, radiating through my body. For hours afterwards, my arm and shoulder nerves protested significantly.

The next day, I examined the ultrasonic cleaner and checked its frequency range. I discovered that it operated in the lightning and piezoelectric range but at a higher power. It functioned at 46 kHz—a slightly higher power range, yet of the same type. My symptoms had the same residual effects as my previous encounters with lightning.

All along, I have believed that the Bosnia Pyramid systems are centred around plasma. Unlike the masculine energy of the Great Pyramids, they generate a feminine (-) energy—Mother Nature sending a grounded stepped leader into the ionosphere, attaching to the masculine (+) positive-charged ions and completing the circuit.

The Bosnian Pyramid of the Sun, in my view, functions as a static ground-based lightning incubator designed to attract plasma from the ionosphere naturally. It is built to capture this magical element of new birth and abundance and circulate it through the waters of Europe and beyond.

THE GIZA PARK PROTOCOL HOW TO START THE SUN

The physical evidence I uncovered was more than sufficient to support my hypothesis regarding numerous systems I discovered—or confirmed to exist—in Bosnia. My confidence increased as collaborative data validated my ideas.

I want to include a summary of the ionosphere.

The ionosphere experiences significant temperature fluctuations, ranging from -99°F at night to 440°F at peak daytime temperatures, with an operational range of approximately 539°F.

It is the ionised layer of Earth's upper atmosphere and acts as a protective shield, as solar radiation strips electrons from atoms in this region.

The Sun is rarely consistent in its temperature range, and many variables influence the energy it emits towards our planet. These emissions first strike the ionosphere—its designated function. One major factor affecting the ionosphere, and consequently the Earth's environment, is solar flares.

Intense M-Class and X-Class solar flares significantly impact specific ionospheric parameters, such as electron temperatures, causing them to exceed 2,000 Kelvin (approximately 3,140°F)—seven times higher than typical. This rapid temperature increase leads to heightened electron ionisation, which, in turn, absorbs the radio frequencies we bounce off its shell.

This phenomenon has caused blackouts in radio communications, including disruptions to the internet. Additionally, increased radiation levels negatively affect the operations of low-orbit satellite systems, such as Musk's Starlink satellites. These extended radiation spikes could have adverse effects on humans and life on Earth.

The influence of the ionosphere on Earth is significant when considering the extreme heat it can reach. I have no doubt that the designers of the Giza Pyramid complex understood the impact of piezoelectric discharge from the Great Pyramid, which occurred every second and influenced the ionosphere's temperature.

Think about it—sudden bursts of high-energy heat alter the ionosphere's dynamics, making it easier for the Bosnian Pyramid of the Sun to attract free plasma ions and draw them to Earth. Remember, the Sahara Desert was once lush and thriving because plasma facilitated new plant growth without requiring traditional topsoil. Plasma represents birth and abundance.

Today, my research and forensic investigations have already concluded that one of the Great Pyramid's primary functions was plasma generation. Secondly, it heated and agitated the plasma river of the ionosphere, causing it to expand and accelerate the accumulation of plasma ions—much like filling a balloon with water. This piezoelectric discharge created the necessary conditions for the Bosnian plasma antenna array to connect with ionospheric frequencies.

Before discussing lightning, let us briefly review plasma, as it is the primary element produced and manipulated by these ancient pyramid systems—similar to modern scientific experiments on Earth today.

Let us also consider ourselves, the Homo sapiens species, and our relationship with plasma.

Human blood consists of 55% plasma, mirroring the 55% clear quartz crystal content of the five 70-ton red granite beams atop the reaction chamber. These beams generate a powerful piezoelectric plasma beam directed into the ionosphere.

Our ionosphere is a Sun-activated, electrically charged protective shield surrounding the Earth. It is a fluctuating environment of highly solar-radiated energy, flowing like a vast river above our heads. Over a normal diurnal cycle, temperatures in the ionosphere fluctuate by approximately 589°F, from 440°F during the day to -200°F at night.

The male sperm is 95% plasma with 5% testicular influence.

As I stated previously, the number one ingredient of our universe is plasma. The International Space Station must electrify an external grid to protect the astronauts when spacewalking because if they don't, they become covered with "plasma jellyfish," as they call them. It is an energy attraction: plasma (+) and humans (-).

Speaking about the speed of scalar/plasma energies, we already understand it is faster than the speed of light. This equates to energy travelling around the Earth in the time it takes to blink your eyes.

Let's compare it biologically, adding to the discussion in point #3 about the male sperm comprising 95% plasma energy.

The techies using our current measurement of data, dealing with our modern computer systems and mobile phones, have even determined the energy of a single male sperm. The male human sperm has 37.5 MB of DNA encoded into it. This means a normal ejaculation represents a data transfer of around 1,587 GB in about three seconds. Sounds impressive.

Now that I have shared this information, I can make my point about the Egyptian god Osiris. If Osiris is the god of fertility, abundance, and rebirth, his so-called tomb, located 108 sacred feet below the desert sands of the desolate Giza Plateau, is protected—like a womb—from some cosmic influences. Reminds me of a CERN design.

I determined that this supposed tomb was not a tomb; it was an ancient dynamic scalar production device. Is Osiris somehow creating plasma and then ejecting this scalar life force energy through the welcoming womb of our Mother Earth?

THE GIZA PARK PROTOCOL HOW TO START THE SUN

If we consider the Great Pyramid's perfect design, we will also see that it is accredited to the great intellect of Thoth/Hermes. From a symbolic perspective, could the Great Pyramid represent the phallus of Osiris? The Great Pyramid was designed to create plasma, and piezoelectrically produced plasmas provide a tremendously powerful electric discharge into the membrane shell of the ionosphere.

The mechanics of this system are perfectly mapped out, creating fertility, abundance, and a suitable environment to resurrect humanity.

If we look at terminating the Ice Age's influence on humankind, we can recognise that once the ice melted away, this allowed freedom of movement globally. It populated every continent on Earth in an amazingly short period—except Antarctica.

Scientists have yet to figure out what caused the two-mile-thick ice sheets to melt quickly, terminating the Ice Age. Their failure to provide a logical explanation culminated in their throwing their hands in the air in confusion and defeat in the 1970s, today known as "The Energy Paradox."

If we look at the influence the termination of the Ice Age had on the humans of that timeframe, it was a vast gift. Humans began to move, populating every continent except Antarctica. Is this bringing with it an environment that supports abundance all over the globe for the benefit of humankind?

The Fertile Crescent of the Mediterranean was created after the ice melted. The Sahara Desert was green, lush, and abundant. New civilisations were created. Our ancestors survived with consistently better environments so we could be here now, thousands of years afterward.

In conclusion, from a mythological and allegorical perspective of my findings, we must consider that maybe the alchemy between

THE GIZA PARK PROTOCOL HOW TO START THE SUN

Isis, Osiris, and Thoth created Horus, the god of the sky—interesting enough.

However, before we close this chapter, I want to add a few more topics we need to discuss before we move on.

First, significant evidence supports the existence of the Bosnian Pyramid of the Sun as an artificial construction and actual, honest science that has been completed. Over 11,000 scientific studies have been implemented and concluded since Dr Semir began his investigations in 2005.

NOTES:

#1) Bosnian POS – Kyiv Institute for Radiocarbon Dating, based on a sample of organic material (fossilised leaves) found between two layers of concrete blocks layered on each other, as I believe were designed as such.

I have theories to introduce later to elaborate more on this topic. Their analysis of the sample provided by Dr Semir concluded that the fossilised leaves were 29,200 years old (+/- 400 years). After a second testing and analysis by his physical investigations, astrophysicist Dr Paul LaViolette determined a more precise dating for the leaves, calibrating them to 34,000 years, making them constructed in 32,000 BC.

This has made the Bosnian Pyramid of the Sun the oldest known structure on Earth, even older than Göbekli Tepe.

That said, I want to continue my thoughts on this topic. Would the Bosnian, Egyptian, and Mexican Teotihuacan pyramid systems be the same age?

#2) The cladding is an artificial conglomerate made from local deposits that were created by postglacial deposits in conglomerate form. Experts examining the crystal content determined that it would have required temperatures exceeding 500 degrees Fahrenheit.

THE GIZA PARK PROTOCOL HOW TO START THE SUN

Studies at the University of Paris found that the concrete samples provided by Dr Semir came from the Pyramid of the Sun. Clinical test equipment determined that the concrete was five times stronger in the megapascal strength test than today's most robust concrete. Ninety-four megapascals reveal a highly compressed and extremely rigid material.

Joseph Davidovits, PhD, performed an electronic microscopic analysis on a sample at the French Institute of Polymers, concluding that its chemical composition is calcium/potassium/geopolymer cement—five times stronger and five times more water-resistant.

The aggregated surface, the hardness of the composite, and the water resistance all support my theories on the function and purpose of this cladding design.

In summary, the polymer used as a binding agent may have been poured over the rubble after it was laid on the pyramid slopes, and the hardening of this product is a process that exists within the pyramids' design itself.

When we speak of stones and concrete that require heat, extreme temperatures sometimes occur, and we cannot imagine how this happens. It is time to discuss lightning.

I'm speaking about a natural ground-to-cloud lightning bolt.

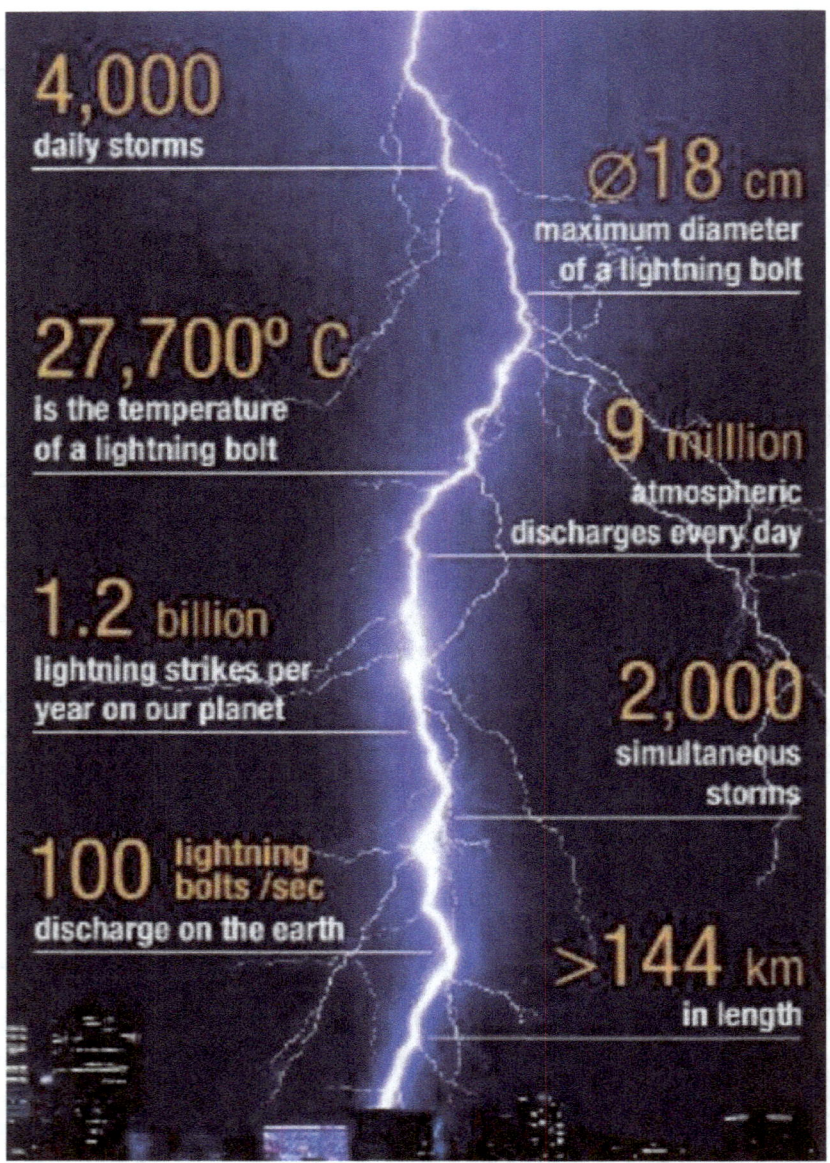

I know more about lightning than most. For example, the lightning you typically see—that big, bright flash, usually loud—comes from the Earth's ground.

Lightning is created in many ways, so let's look at a couple. First, you're dealing with clouds containing layers of positive (+)

THE GIZA PARK PROTOCOL HOW TO START THE SUN

and negative (-) energy—clusters or pockets of energy constantly shooting back and forth at each other.

Often, they send out step leaders, as they are called. Sometimes, these are negative, energetic, and always searching for their opposite. Remember, the positive is always attracted to the negative—the chase is on all the time.

See images:

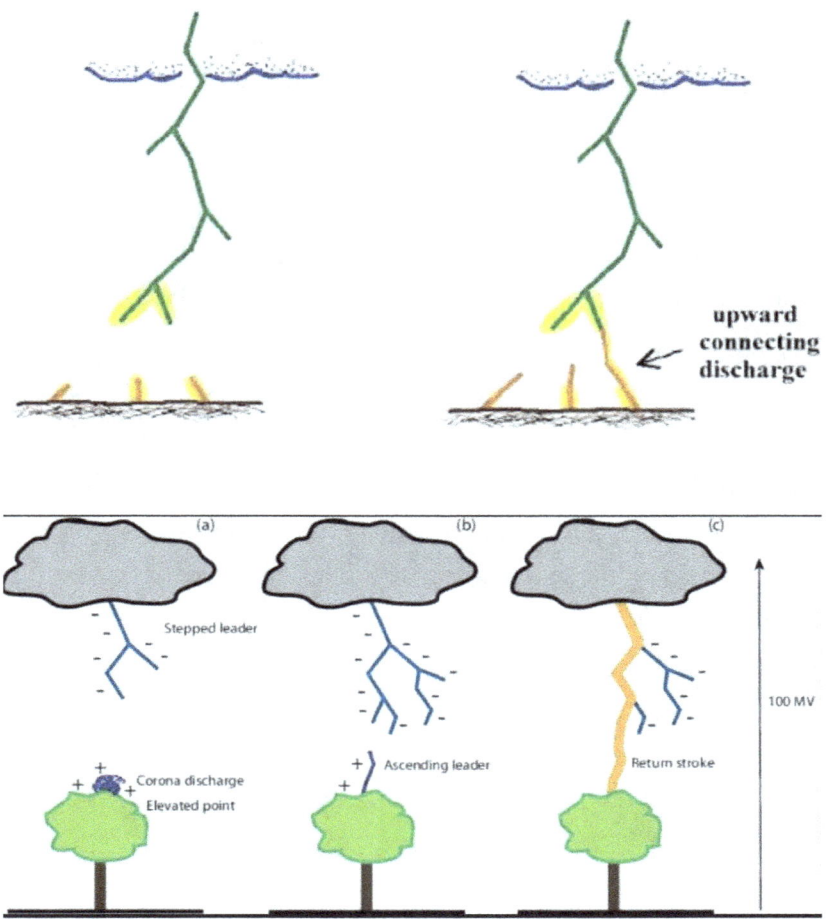

THE GIZA PARK PROTOCOL HOW TO START THE SUN

When you see the brilliant flash and sometimes hear a boom, they originate from the Earth—just like what occurs at the Bosnian Pyramid of the Sun.

If the Pyramid of the Sun acts as a lightning incubator, with immense stone weight and pressure on top, it will significantly influence the interior components, such as the stones inside. The intense heat coursing through them—five times hotter than the surface of the Sun—will substantially alter their physical properties.

The surface of our Sun is 10,000°F, which is incredibly hot. However, lightning reaches 50,000°F—five times the temperature of the Sun's surface.

When you combine this extreme heat, even for a short period of just two or three seconds, with the tremendous weight and pressure in the environment, the heat exchange happens so rapidly that little radiant heat energy can be measured in the surrounding geology.

Under these conditions, rocks and conglomerates can be influenced without requiring a prolonged heating sequence sufficient for the stone to reach a melting state and mix different composites.

Barbara Tosti and Jesper Johansen provided images of stones they retrieved while excavating the healing tunnels beneath the Bosnian Pyramid of the Sun.

THE GIZA PARK PROTOCOL HOW TO START THE SUN

THE GIZA PARK PROTOCOL HOW TO START THE SUN

THE GIZA PARK PROTOCOL HOW TO START THE SUN

The following images are physical examples of lightning shooting out of the sand and petrifying the sand, capturing the lightning bolt's physical characteristics.

THE GIZA PARK PROTOCOL HOW TO START THE SUN

The Fulgurite image below is from my research project on a channel harbor dredging operation and beach reclamation project. It took me three years to complete, as the beach eroded 10 feet tall.

As you can see, it was simple to take the fulgurite fragments and create a natural lightning image using petrified sand from Lightning.

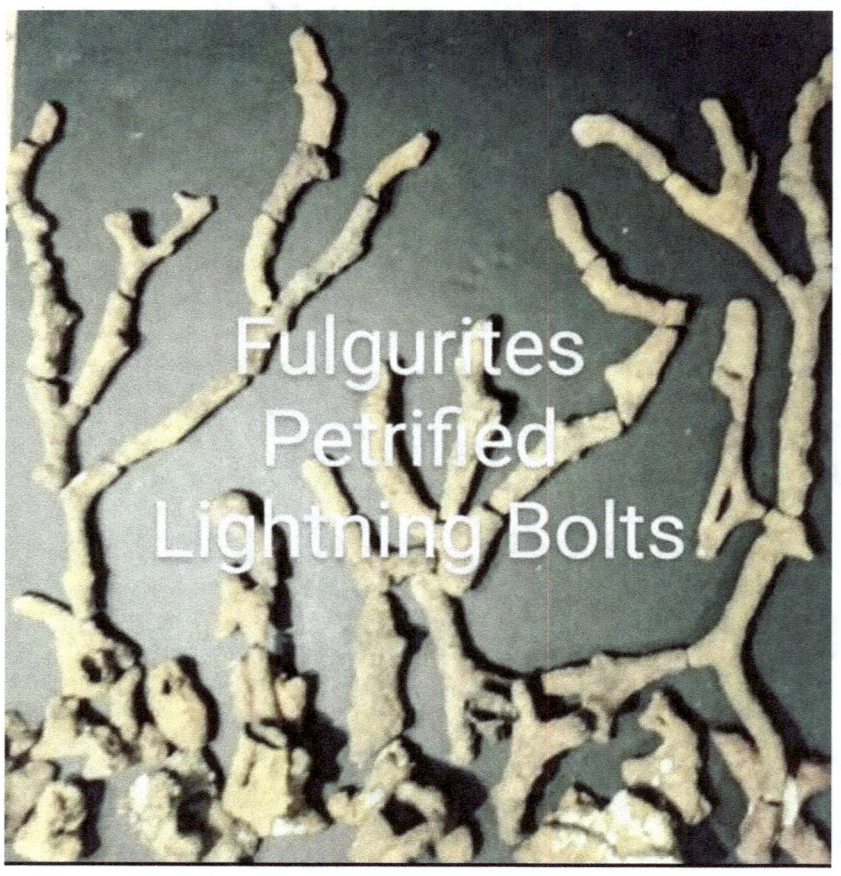

The following images came from the same beach reclamation project and harbor dredging operations. These two images perfectly exemplify the power of a 50,000-degree F temperature moving at scaler speeds faster than light.

THE GIZA PARK PROTOCOL HOW TO START THE SUN

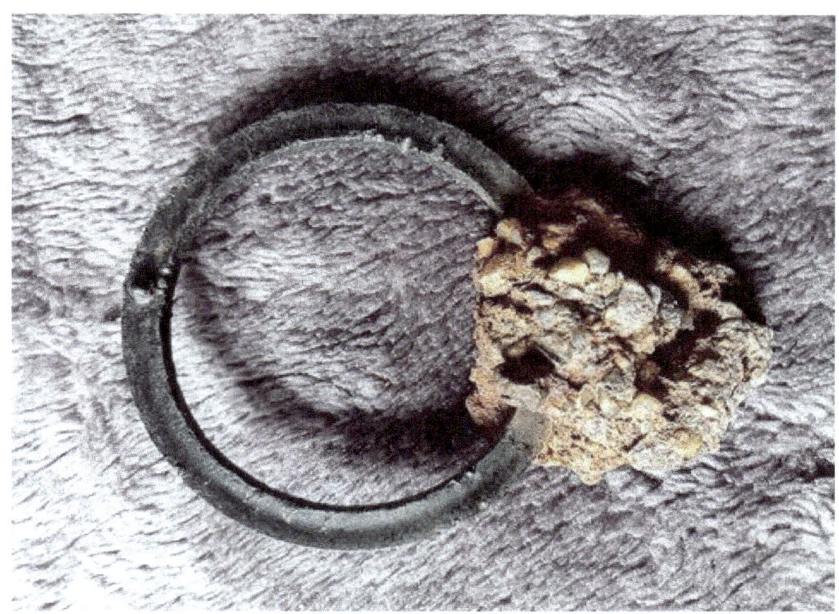

This image is a rare find, featuring a joint plastic gasket commonly used by commercial watermen in their sea strainers and filters. These components channel cooling seawater through a vessel's engine to prevent overheating, as boats do not have radiator systems like automobiles. Instead, they circulate raw seawater through a heat exchanger. Due to debris in the water, sea strainers require frequent cleaning—often daily. Over time, these gaskets tend to leak, and most watermen simply discard the old ones overboard and replace them with new ones.

It is remarkable to consider that a lightning bolt, heating sand to 50,000°F, could fuse the sand together so rapidly that it bonded into stone—yet the heat was so brief that it didn't melt the plastic gasket. Instead, the surrounding stone fused around it. All of this occurred in seawater, which acted as both an electrolyte and a cooling agent.

Now, imagine immense pressure bearing down on different materials while extreme temperature and voltage surge through them repeatedly. Over time, this process naturally leads to what I call "stone fusion." It is a normal and natural event on Earth, though few truly understand how lightning operates.

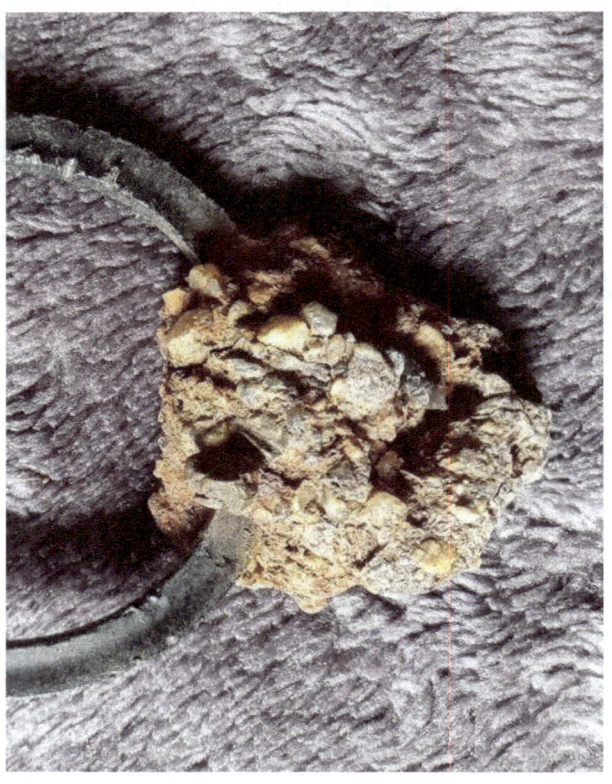

THE GIZA PARK PROTOCOL HOW TO START THE SUN

Also, Horus, the falcon-headed god, soars in the sky and is worshipped at noon in Egypt as Ra, the Sun God/Horus.

Nikola Tesla learned much about plasma and developed his famous 369 formula, which was everything he needed to understand when working with the Germans at the Giza Plateau before the war. The knowledge he gained led to the development and design of the Experimental Model, which he constructed while working with lightning (plasma) in Colorado. He then resolved to build a working model at his famous Walden Cliff Tower Project on Long Island, New York.

Walden Cliff mimics the systems upon the Giza Plateau with a high-powered modern design created by Nikola Tesla. Once Tesla tested its operations and made determinations based on physical assessments, he updated his design with the intention of selling the system to every major country on Earth.

His first sales presentation, offering his designed Ionospheric Heater system to the US, failed. The government declined his offer, claiming his requested one-million-dollar price tag was too expensive. Tesla then approached the Soviet Union and made a successful sale, making them the first in the world to construct a working version of an Ionospheric Heater.

Once the Soviet system was completed and recognised by the US government, they rescheduled another meeting with Tesla to reconsider a purchase. At the time, Tesla was recovering from a near-fatal hit-and-run automobile accident as he left a meeting with J.P. Morgan. The US government, having learned that the Soviets had constructed the system, sought a second look at Tesla's design.

However, this meeting never took place. Tesla mysteriously died the night before the scheduled appointment at his hotel. George Bush Senior was assigned to oversee the investigation into Tesla's death.

During his investigation, it was reported that the US government confiscated 75 trunk containers filled with Tesla's genius. These trunks were immediately shipped to the Pentagon for thorough inspection under the supervision of MIT Professor John G. Trump, the paternal uncle of President Donald Trump, who was tasked with reviewing all the technical information inside.

John G. Trump was an electrical engineer, physicist, and inventor of rotational radiation therapy. He created the first million-volt X-ray generators, which were estimated to be worth between $1,000,000 and $8,000,000 at the time of his death.

One characteristic for which Tesla was notorious was his love for super-high power. A costly example of this was when Tesla was conducting experiments with lightning in Colorado, transmitting powerful voltages into the Earth. Some miles from his laboratory stood an AC electrical generator facility that provided electricity to local communities. During one of Tesla's experiments, a rogue super lightning bolt destroyed some of the giant-sized electric generators at the power plant.

Eventually, Professor Trump isolated the information needed to construct the same system Tesla had sold to the Soviets. The Massachusetts Institute of Technology (MIT) then designed a working model, which eventually led to the construction of the

THE GIZA PARK PROTOCOL HOW TO START THE SUN

HAARP Ionospheric Heater System in Fairbanks, Alaska, in 1958. HAARP stands for High-Frequency Auroral Research Program. The project was funded by the US Air Force, the US Navy, and the Defense Advanced Research Projects Agency (DARPA).

Today, my research and interviews with those installing these systems globally suggest that an estimated 250 of these similarly designed systems exist worldwide. Massive versions have even been built on large barges that are towed with tugboats. I know for a fact that one of these docks in Hawaii.

Let us continue our discussion on the analogy of the Flow Chart systems.

Let's return to 7,000 BC on the summer solstice and initiate the Weather and Climate Modification Systems.

On day 70, Sirius was absent, and the crews were preparing for this early sunrise event.

THE GIZA PARK PROTOCOL HOW TO START THE SUN

Giza Start-Up Manual

Section I

The Great Pyramid component of the Giza Park Industrial System consists of FIVE (5) dedicated operating systems.

These **FIVE** Major systems are:

SYSTEM ONE----- Pressurized Fuel supply system, Liquid Fuel Pumping System

SYSTEM TWO----Hydrogen Disassociation Chamber System

THE GIZA PARK PROTOCOL HOW TO START THE SUN

SYSTEM THREE-Chemical and Product Management System

SYSTEM FOUR---Side Draft Fuel/Product Separator System

SYSTEM FIVE-----Primary Reaction Chamber

Giza's systems' requirements are based on strict adherence to critical guidelines.

FUEL REQUIREMENTS:

Ignition Starting Fuel: Chemically induced hydrogen gas.

Operating Fuel: Lake Moeris seawater aqueducts, containing six ion-rich minerals. The water has been heavily structured and disassociated by chemicals, minerals, and kinetic energies to obtain the hydrogen gas needed for plasma extraction.

Note: In writing this manual, I have organised and chronicled logical and sound operating procedures based on 40 years of experience, adhering to industry standards.

The *Giza Start-Up Manual* is the first of its kind. It contains comprehensive instructions on how to start and operate the Egypt Weather and Climate Modification Systems located in the Sahara Desert region of the Giza Plateau.

THE GIZA PARK PROTOCOL HOW TO START THE SUN

At this stage of the process, all pre-start tasks are as follows:

The Chemical Injection Teams are on-site and ready. The Chemical Injection Crew has prepared the appropriate amounts of two chemical compounds necessary to initiate the system's 295-day operation period.

Crews are stationed at assigned locations along the Nile, from Hawara to 65 miles north of Abu Rawash. The other four vessels of

THE GIZA PARK PROTOCOL HOW TO START THE SUN

the stored fleet are afloat on the Nile River at designated troublesome areas that require close supervision during this compelling and dangerous event—an achievement modern man has yet to accomplish!

Filler up, please.

Stage 1

Two chemical teams are on-site. The southern chem team injects hydrochloric acid into the southern injection tube of the *Start-Up Ignition System/Water Disassociation Chamber.* Simultaneously, the northern chem team injects the second chemical compound, hydrated zinc chloride, into the northern injection tube.

Both chemical injection tubes feed directly into the *Start-Up Chamber.* Once the teams are confident in the setup, all systems and crews prepare for the start-up sequence.

The Lead Controller gives the command to sound the alarm, notifying the chem teams that the production of hydrogen gas is about to begin.

The chemical reactions from mixing these two compounds utilise the *Ignition Chamber's* chemical repository, producing hydrogen gas. At this stage, the *Ignition Chamber* can now be considered a *Hydrogen Gas Incubator.*

Observing the structure, the small square hole in the wall serves as the chemical fill shaft or tube, which releases its contents onto the granite floor, where they mix with the chemical entering from the opposite shaft. The larger hole allows fuel to enter the *Ignition Chamber,* while the corbelled archway is specifically designed to facilitate hydrogen and oxygen disassociation once the system is operational.

THE GIZA PARK PROTOCOL HOW TO START THE SUN

The chemically induced, low-ignition-point hydrogen gas becomes the starting fluid necessary to initially "fire up" the Great Pyramid in one of the most magnificent events ever witnessed on planet Earth. It is a design so magnificently and beautifully developed that it creates an environment allowing humans to thrive with abundance and opportunity anywhere on Earth—except Antarctica.

The carefully metered chemical hydrogen gas production continues accumulating inside, filling the entire interior, including the five major systems of the Giza Pyramid. As atmospheric pressure shifts, much of the manufactured hydrogen gas is forced through the circulating return line inside the Ignition Chamber.

Once the hydrogen gas enters the shaft, it descends into the Subterranean Chamber and flows freely through the static "Cadman Pump." The gas moves counterclockwise, filling the vertical descending passageway as it rises into the corresponding ascending passageway.

Hydrogen gas is highly flammable and volatile under certain conditions, which we will explain as we go. Being lighter than air, it fills the Ignition Chamber quickly. As the chemicals continue to flow into the Ignition Chamber, the volume of hydrogen gas becomes denser than the resident oxygen.

Hydrogen gas then migrates from the Ignition Chamber into the elevated return drain intake. The intake pipe leads directly into the suction side of the Cadman Pump, positioned directly below it. Once the heavily saturated hydrogen gas has purged through the dry and static Cadman Pump, the escaping gas is whisked away like smoke from a fireplace chimney, drafting the explosive fuel up the Descending Passageway. From here, the gas spills into the Ascending Passage, accelerating as the nearly vertical passage expands, flooding the enormous hollow void of the Grand Gallery with the ultra-light, explosive gas.

The injection of chemicals continues until hydrogen gas saturates all the interior compartments of Giza, forcing the oxygenated air out of the pyramid through the two vents inside the Reaction Chamber.

Scrutinised Priority: The chemical teams understand the precise proportion of chemicals to be injected into the Ignition Start-Up Chamber. The volume of chemicals introduced by gravity will be carefully dispensed. The hydrogen gas reaction will be violent and deadly—being nearby would be ridiculously foolish.

THE GIZA PARK PROTOCOL HOW TO START THE SUN

As soon as the manufactured hydrogen fills the Ignition Chamber, it begins to seep into the Grand Gallery. Now, for the first time, one of the Grand Gallery's design functions becomes evident—it is perfectly engineered to contain massive amounts of hydrogen gas.

As the Grand Gallery fills from the top down, the hydrogen gas seeps into the Ante Chamber and subsequently into the Reaction Chamber, which has two 8-inch vents exposed to the outside air. This creates a chimney effect as the Reaction Chamber reaches its maximum capacity.

At this point, hydrogen gas is filling the airspace surrounding the Great Pyramid. As the interior becomes saturated, newly manufactured gases are forced deeper into the Descending Passageway and the Grotto well, eventually reaching the bottom where the Cadman Pump operates with its Egyptian "Ligman Vita" industrial marine specialty wood pump mechanism, generating a commotion 60 times per minute. However, today is particularly significant, as the start-up sequence requires perfect coordination and timing.

Once total hydrogen gas saturation is achieved, the observation chemical teams notify the liaison officer.

Stage 2

Standing by, the crew inside the park and those scattered along the Nile are notified to prepare. Horns sound, fires are lit, and explosive signals are detonated. Everyone remains on high alert beneath the cover of darkness. Dawn approaches, and excitement fills the air.

Seemingly in mere moments, Sirius is spotted cresting the horizon at 114 degrees east.

Immediately, the skilled boatmen lift the heavy granite stone to control the tunnel entrance. This intake opening triggers a powerful vacuum effect, inhaling the fuel with tremendous force. The resulting suction causes the fuel to cascade like a waterfall, exploding into the primed and ready Cadman Pump.

See the image of the massive Cadman Pump housing in the Subterranean Chamber below.

Like none on Earth before, the Magnificent Super Fuel rushed through the underground tunnels and plummeted into the Subterranean Chamber, one hundred feet directly below the Great Pyramid's frame. As the fuel gained momentum, the kinetic energy of water and its dynamics became a powerful force—water torrents surged as they reached the first Giza system, located in the Subterranean Chamber. Cadman's Pump is the heart of Giza.

A violent marriage of 80 tons of water was sucked almost 100 feet straight down a narrow tunnel, only to impact the Clapper/Hammer Valve of the Cadman Pump.

Remember our discussions on piezoelectricity and how electricity is created using mechanical energy? Here, it is a dynamic hydraulic water ram.

The Clapper/Hammer is the only mechanical component of the Cadman Pump itself. However, four mechanical components exist

THE GIZA PARK PROTOCOL HOW TO START THE SUN

within the Great Pyramid's fuel supply system, which I will provide more details about as we progress.

The Clapper/Hammer component has two designed functions. The term "Clapper" relates to its function within multiple single-displacement pumps. "Hammer" refers to the mechanical force it creates, which impacts the Check Stone with sufficient energy to resonate through the "Reaction Chamber" above it—an area specifically designed for this purpose.

Inside the Reaction Chamber, we have already detailed the five 70-ton red granite beams in the overhead structure, each containing 55% clear quartz crystal, waiting to discharge.

However, we have yet to discuss another crucial component, which plays an essential role in the ignition start-up stage.

The Piezoelectric Direct Discharge/Plasma Transformer also serves a dual function, as is often the case with ancient systems of this incredible design.

The image below reflects a grossly disfigured, rectangular-shaped red granite box containing 55% clear quartz crystal.

THE GIZA PARK PROTOCOL HOW TO START THE SUN

This particular grade of stone is tough, and you can easily identify the massive amount of what appears to be severe abrasive erosion of granite and clear quartz crystals.

My conclusion, as a hint to where our travels are headed, is this:

This Egypt Weather & Climate Modification System has operated since 32,000 BC in two controlled cycles, totalling 27,100 years. We are witnessing damage to this hard red granite from at least 27,100 years of piezoelectric discharges.

Observe this device's sharp edges—every one of them has been eroded, as this is where the piezoelectric plasma discharges are designed to escape from the stone.

The following image reveals physical evidence of extremely high heat damage: the discolouration of the red granite Reaction Chamber walls. You can easily see the damage to the walls; however, it is much more severe around the rectangular Ignition Box.

THE GIZA PARK PROTOCOL HOW TO START THE SUN

If I were conducting a typical structural forensic fire investigation, I would declare this location the ignition source.

Now, let's imagine the Great Pyramid saturated with a highly explosive gas inside every crack and crevice—a time bomb awaiting detonation from the slightest spark.

Let me describe the interior a little more so you can understand the arrangements and the complex dynamics that will soon take place.

Beginning from the Cadman Pump Room, 100 feet below the ground, let's take a quick tour of the inside of the Great Pyramid.

Entering the Descending Passageway, we begin a steep elevation until we encounter another tunnel that changes our direction, turning 90 degrees into the Ascending Passageway. As we climb this incline, we reach an area with an intersecting junction of passageways and tunnels near the end of the Ascending Passageway. To the left is the smaller, horizontal entrance passage into the Ignition Start-up Chamber.

Looking in that direction, we would see the second mechanical device, known as a check valve, which is also a clapper design—one moving part sometimes called a flapper valve.

Straight ahead is the entrance to the hollow rows of corbels as we merge into the Grand Gallery and Passageway.

The third mechanical device is a second check valve, identical in design to the first but much larger due to its location outside the Reaction Chamber. It is situated at the top of the Grand Gallery and is known as the Great Step. This check valve seat is where the clapper valve rests. It accommodates a flow of fuel in two directions—one fuel enters and exits after becoming electrolysis plasma-infused from the super piezoelectric ultrasonic 28 kHz energy.

THE GIZA PARK PROTOCOL HOW TO START THE SUN

After climbing the Great Step and entering the Ante Chamber, its new designation will be the Venturi Side Draft Vacuum Carburettor.

After crawling through the narrow entrance to the Venturi, we stand up and can see rows of grooves carved into the stone, designed to house the last and fourth mechanical components of the Great Pyramid's fuel system inside the Venturi. This small area would contain a floating device that moves vertically and operates in sync with the other systems as an atmospheric balancing component.

For example, fuel management of two fuels controls the air-to-fuel ratios due to two vents in the Reaction Chamber.

A note: For the Great Pyramid's fuel systems to operate perpetually and synchronically, these two check valves (flapper-type) must be installed, allowing pump circulation in only one direction.

While we are at this location, I want to explain an exciting and brilliant design that few are aware of—something Chris Dunn pointed out to me. The stone platform, flooring, and support for the Reaction Chamber contain a large void beneath them filled with silica crystals (sand). The floor is designed to capture the resonance of the Cadman Pump, creating vibrations in the floor itself. The only thing sitting on it is the ignition piezoelectric device.

THE GIZA PARK PROTOCOL HOW TO START THE SUN

So now we have all the components needed to explain the ignition dynamics in detail.

What do we know?

The Great Pyramid is filled with the maximum amount of hydrogen gas in preparation for ignition.

THE GIZA PARK PROTOCOL HOW TO START THE SUN

This hydrogen gas is a chemically created compound that results in hydrogen gas. Once this gas is used for the ignition process, the future hydrogen gas produced will come from the dissociation of H_2O molecules contained within the liquid fuel medium.

This incredible, super-energised, and powerful liquid fuel is subjected to gravity, vacuum, and water force.

The moving waters rapidly adhere to the designed counterclockwise circulation of the pyramid system.

Remember, the pump is static and non-operational, and the entire pyramid has already been filled with high-density hydrogen gas. However, the gas offers little resistance to the immense power and force of the denser liquid fuel. As the hydrogen gas is squeezed out of the lower compartments, it rises upwards to the top of Giza's interior, accumulating in the reaction chamber.

I mention this now because the power of the rushing water is so great that the force produces what is known as a **hydraulic ram**. This hydraulic ram acts like a piston or a trigger on a grill lighter.

However, this piston is not made of metal, like those in automobile engines or lawnmowers. Instead, it is formed from water. Spiralling uncontrollably, this hydraulic piston further compresses the already rising hydrogen gas. As the gas encounters greater pressure and resistance, it becomes denser, and the hydrogen molecules grow increasingly unstable as they are crushed against the piezoelectric granite ceiling by the immense force of the water.

Liquid alone has a specific gravity and is denser than hydrogen gas. When combined with the dynamic force of water, it creates the perfect environment to produce ignition unlike anything else on Earth. This is why the Great Pyramid will always be the greatest—because of its extraordinary diversity.

THE GIZA PARK PROTOCOL HOW TO START THE SUN

Stage 3

This is how the hydrogen gas is ignited.

One major factor in the ignition process is timing. Many powerful universal cyclic events co-occur around Earth in this precise location. We are dealing with a concise duration of only a few minutes at most.

The force of the water alone causes hydrogen gas to be compressed against the ceiling and walls of the Reaction Chamber, which is identical to an engine's combustion chamber. The immeasurable force generated from the impact of the induced hydraulic ram of liquid fuel becomes the "trigger" mechanism.

In a nanosecond, the resonance mechanical energy and the compressed hydrogen gas explode violently as millions of volts of current from the quartz crystal granite are released from the ignition device in the Reaction Chamber. This powerful Reaction Chamber also contains the power grid, comprising four 70-ton piezoelectric granite blocks and over 200 tons of clear quartz crystal piezoelectric energy (plasma).

These piezoelectric discharges from humankind's most tremendously powerful energy device simultaneously fill the Reaction Chamber with ample plasma for water dissociation, creating more plasma as the main release impacts the ionosphere. The hydraulic ram is comparable to a combustion engine. The ignition device in the Reaction Chamber acts like a spark plug, and the energy inside the Reaction Chamber becomes the ignition source—the piezoelectric plasma spark.

Boom!

The combustion of the hydrogen gas explodes.

We have ignition. With the energy and destructive force of an atomic bomb, this violent power thrusts against the hydraulic piston of liquid fuel, forcing it downward in the opposite direction of its entrance.

As a reminder, we previously discussed that the check valve design prevents back pressure and guides the water counterclockwise. We know that the combustion energy transfer cannot backtrack, reversing the counterclockwise circulation inside the pyramid systems and returning to the Cadman Pump.

So, what happens now?

We have learned that the Reaction Chamber has two vents. These vents help balance the fuel/air ratios and act as an exhaust system. During this start-up process, the first catastrophic combustion would have blown a stream of liquid fuel great distances north and south. However, due to the small, restrictive size of the vents, most of the energy transfer is pushed downward. This means the hydraulic ram piston reacts to the explosive force, causing a temporary reversal of the liquid fuel direction.

Travelling through the water system, the first ram of energy moves down through the Grotto-Well feed pipe, located on the floor adjacent to the Venturi Intake. From the Venturi Intake, the energy moves through the plumbing system until it reaches ground level at the Grotto. The Grotto is already filled with liquid fuel from the force of the ignition.

A deep shaft runs into the Earth from beneath the Grotto. This Grotto well shaft functions as a conduit, flooding its temporary volumes of super fuel into the discharge side of the Cadman Pump. The Cadman Pump, which had been idle, suddenly receives this gigantic shock wave of energy from the mighty hydraulic ram. The explosive back force is too great for the once-resting Cadman Pump to remain inactive, and within a short period, it becomes primed and operates simultaneously and sequentially.

THE GIZA PARK PROTOCOL HOW TO START THE SUN

Once the Cadman Pump starts, Giza's heart begins pumping, circulating 80 tons of saltwater and reaching tremendous pressures of 3,200 PSI. From then on, Giza operates as a perpetual machine, performing continuously with routine scheduled maintenance for many lifetimes or centuries.

Before continuing, we must realise that the Reaction Chamber floor rests atop a platform designed to vibrate. Beneath this vibrating platform, a hollow cavity filled with sand (silica crystals) provides a vibrational resonance chamber. This setup is perfect for capturing and boosting the intensity of the pump vibrations, creating mechanical energy.

The expansive sonic vibration produced by the massive Cadman Pump travels through the dense granite pyramid blocks, accurately targeting the resonating chamber below the Reaction Chamber. These sonic vibrations are captured by this ancient subwoofer, vibrating the floating platform that supports the Reaction Chamber. Like music to our souls, every stroke of Giza's mighty heart pounds the drum 60 times a minute—the same as ours. An interesting coincidence.

This concludes Giza's start-up procedures.

Once the Lake Moeris super fuel seawater has undergone all the processes inside the Great Pyramid, pump pressure and dynamic flow propel the liquid at a constant rate out of the Great Pyramid. Once it exits, the treated seawater is first sent to the Tomb of the Birds.

#10) Tomb of the Birds – This water structuring distribution facility uses the same methods, minerals, and resonance to structure the water before integrating it with the rest of the industrial system. While smaller in scale than the Serapeum, the newly treated and structured seawater is now prepared and ready for distribution alongside the other two pyramid systems and beyond. The water is now sent to a tunnel distribution system under pressure.

THE GIZA PARK PROTOCOL HOW TO START THE SUN

#11) Tunnel Distribution System – The prepared seawater is pumped eastward via:

Tunnel (A): Directly to the Ionic Plasma Reflux Pyramid for water treatment, including the infusion of captured ionospheric reflux plasma ions that travel to Earth with each piezoelectric discharge from the top of the Great Pyramid.

Tunnel (B): To the deepest tunnel on the Giza Plateau, which runs beneath the lowest level of Osiris's Tomb. Three drains beneath the so-called tomb capture the treated water, which then moves eastward towards the Geyser Sphinx. I will elaborate more on this as we include the other systems.

Tunnel (C): To the south of the Electric Pyramid for further water structuring. The unique cladding design of this structure differs from the other two pyramids. The bottom half is covered in a high concentration of quartz crystal granite, while the top half is clad in white limestone, like the other pyramids. This design creates a substantial piezoelectric discharge, producing a lightning bolt effect every second the Cadman Pump strokes.

#12) Ionic Plasma Reflux Pyramid – This pyramid is the focal point for the ionosphere's earthbound direction, as it releases hot positive ions from the breached ionospheric membrane, stimulated by the Great Pyramid's piezoelectric activity. Acting as a heat sink with a negative polarity, it adheres to the law that positive charges are always attracted to negative ones.

Many researchers and archaeologists believe the current Plasma Flux Pyramid is a reconstruction. Examining the existing, significant, ground-level exposed limestone bedrock pad where the Ionic Flux Pyramid was constructed reveals a larger foundation than the other two pyramids combined. This suggests a more extensive structure previously stood there.

THE GIZA PARK PROTOCOL HOW TO START THE SUN

A thorough understanding of these systems and their functions has led me to strongly disagree with the reconstruction theory. Instead, I believe this vast, oversized foundation is a natural bedrock limestone heat sink that functions as a ground (-) negative waveguide antenna, attracting supercharged (+) positive ionospheric plasma flux. The logical purpose of this is to capture high-energy ionospheric plasma flux, which becomes a hitchhiker, travelling to Earth, attached to the piezoelectric (-) negative discharge from the Great Pyramid. As the piezoelectric discharge beam recoils from the ionosphere, the plasma is drawn towards the Ionic Plasma Reflux Pyramid.

As previously stated, plasma is scalar energy that travels through limestone at the speed of light. The size of this antenna grid, combined with the capacity of the Ionic Plasma Reflux Pyramid, enables it to capture and manipulate an amount of energy that modern scientists cannot even estimate.

This incredibly high-powered (-) negative beam attracts the ionospheric plasma (+) positive charge, aggressively facilitating diffusion and photon absorption. This action influences the water flowing through the Plasma Reflux Pyramid structure, infusing plasma-positive ions into the prepared water. The water then exits the structuring chamber on the east side of the pyramid and cascades down the open-air causeway.

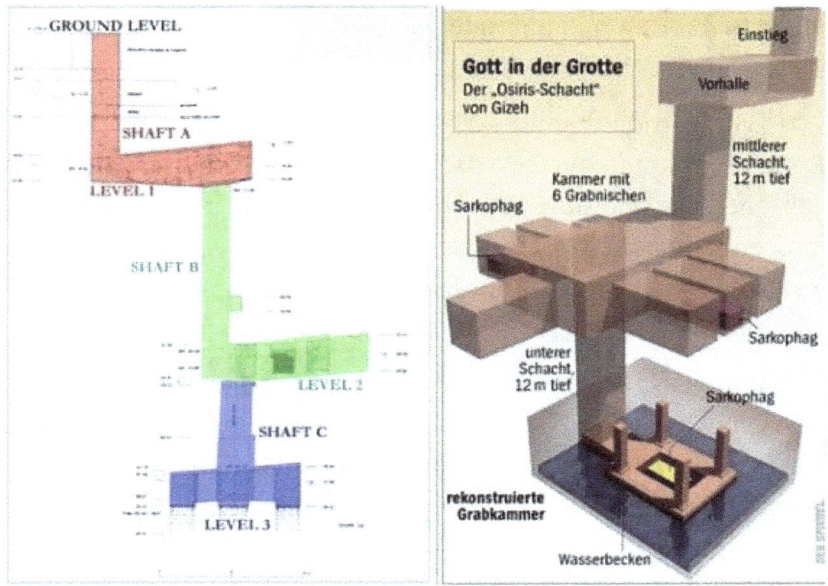

As the water flows a short distance down the causeway, a metered amount of this plasma-infused water enters a drain-like opening in the ground, supplying the necessary Plasma (+) Ions into the scalar system of Osiris's Tomb.

This plasma-infused water, after cascading down the causeway, passes through two water-structuring chambers before being directed into another underground tunnel, dropping 116 feet below ground level (BGL). This tunnel then turns northeastward and connects at a junction to Tunnel (B), which I previously described as the deepest tunnel. It runs beneath Osiris's Tomb and ultimately leads to the Geyser Sphinx.

However, the highest volume of water that is not siphoned off into the Osiris Shaft—this highly charged, structured, plasma-infused water—continues flowing down the lengthy causeway. Along the way, it gains additional kinetic energy and absorbs solar photons. Upon reaching the end of the causeway, in front of the Geyser Sphinx, it enters the second Plasma Reflux Pyramid Water Structuring Antechamber. As mentioned earlier, once the water circulates through this intricate maze of structuring and energising minerals, it exits into the Mixing Basin.

#13) Electric Pyramid—The Electric Pyramid is the smallest of the three pyramids on the Giza Plateau. It lies the farthest to the south and, like the other two pyramids, has a unique design and purpose. Unlike the Great Pyramid's White Tura Limestone casing stones, this pyramid's exterior features a different material and lacks a capstone.

The Electric Pyramid differs from the second-largest pyramid, as it has a single bottom row of rose quartz granite stones on the exterior, followed by white Tura Limestone casing stones, and a capstone.

On the other hand, from the ground level up to halfway to its peak, the Electric Pyramid's capstone is composed of rose quartz granite. I will provide more details on the significance of these differences as we progress.

This system includes three mineral water-structuring antechambers and a long causeway above ground. The water is exposed to solar photons and receives an electric current into the seas due to the piezoelectric discharge from the quartz granite, triggered every second by the Cadman Pump's stroke. This scatter of piezoelectric current not only dissipates the stray hydrogen gases produced during system operation but also discharges electricity from this massive quartz structure, mimicking the effect of a lightning strike.

Lightning and piezoelectric currents create an electrolytic cell, which utilises electric currents to break down substances into simpler forms. Electrolysis is a standard method used to produce hydrogen and oxygen from water. In this case, most of the oxygen escapes into the external environment, while hydrogen becomes an abundant gas. The plasma ions in the water originate from this

THE GIZA PARK PROTOCOL HOW TO START THE SUN

medium, impregnating the treated and structured seawater flowing down the lengthy causeway to create plasma within the water.

This system also incorporates three mineral water-structuring antechambers and a long above-ground causeway. The water is exposed to solar photons and receives an electric current while flowing through the river, mimicking the Sun's influence on the Ocean of Ionospheric Plasma circulating above our heads.

The piezoelectric discharge from the quartz granite occurs with every stroke of the Cadman Pump. This discharge not only dissipates stray hydrogen gases but also generates additional plasma ions within the water, mirroring the effects of lightning strikes.

Let us now continue and bring all these elements together after we detail two essential systems: Osiris's Tomb and the Geyser Sphinx.

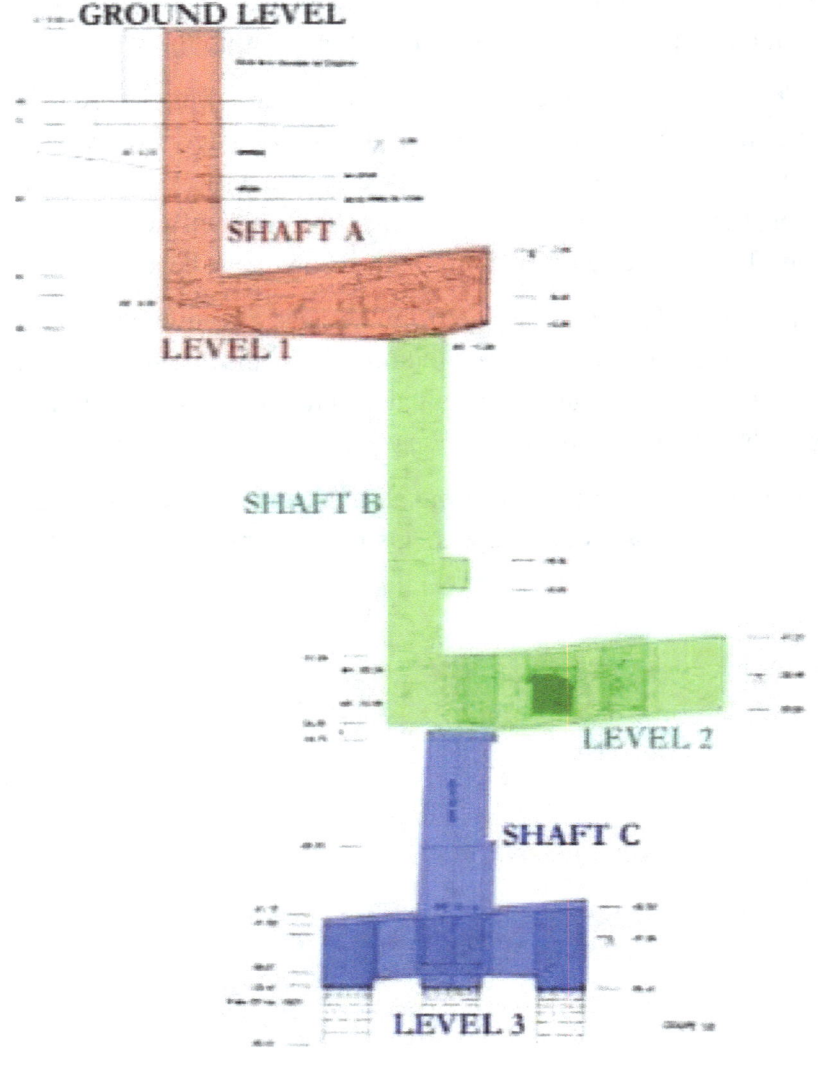

#14) Osiris Tomb – This is one of the most mysterious, complex, and least understood systems on the Giza Plateau—or, should I say, beneath the Giza Plateau. It comprises three levels after entering the intake tunnel from the Ionic Pyramid's causeway.

THE GIZA PARK PROTOCOL HOW TO START THE SUN

Level (a) is 9.6 metres (35 feet) below ground level (BGL). Level (b) descends a further 13.25 metres (43.5 feet), now reaching a depth of 78.5 feet BGL. From there, it continues downward to level (c), which is 7.5 metres (24.6 feet) deeper, bringing the total depth to 103 feet BGL. As the treated water flows down the tunnel to the very bottom of the so-called Osiris Tomb, the chamber beneath floor level contains the sarcophagus-like stone, now reaching a sacred measurement of 33.07 metres (108 feet) BGL.

Earlier, I identified the lowest-level tunnel from the Tunnel Distribution System as Tunnel (B), which is 35.45 metres (116 feet) BGL. It is named Osiris's Tomb, the resting place of the god of the underworld, Osiris.

At this location, I made one of my most significant discoveries. Once I had fragmented and identified each mechanism, reverse engineering enabled me to prepare the rest of the systems analysis, leading to my final determination.

I named Osiris Tomb's system a **Quantum Superconductive Plasma Electromagnetic Cyclotron**. It is designed to send long-range plasma-positive ions into the solid limestone bedrock. An interesting note: the limestone bedrock has a negative polarity and acts like a sponge, enthusiastically absorbing these powerful positive plasma ions.

In these circumstances, the quantum difference between negative and positive ions is that negative ions can only travel a short distance, whereas positive ions are unlimited and can travel indefinitely in the right environment.

This device generates a powerful scalar energy that is quantumly directed into the Earth—more details to follow.

Referring to my earlier description of Tunnel (B), it exits from the Water Distribution System—the deepest of them all—and travels east towards the Geyser Sphinx. Due to the design of the third level of the Osiris Tomb, which is not aligned with the four cardinal directions (N/S/E/W), the tunnel is angled eastward and must cross diagonally beneath the Osiris Tomb.

As this pressurised water flows through, it captures the treated seawater from Osiris Tomb, combining the ingredients and structural properties of both systems before transporting them towards the Geyser Sphinx.

A few hundred feet from the Geyser Sphinx System, a powerful stream of water arrives from the completed treatment cycle of the Electric Pyramid System, as I mentioned earlier.

The blending of water from Osiris's Tomb, the Ionic Pyramid, and the Electric Pyramid—along with every treatment system involved—is perfectly formulated before being channelled through the Geyser Sphinx.

I love the image of a man standing inside the head of the Sphinx. Observe the size of the hole and the severe erosion surrounding it, which has influenced the entire head.

This physical geological evidence suggests that water once flowed from the head with considerable velocity.

While on the topic, the reason the head of the Sphinx is so significantly smaller than its body is a direct result of physical erosion.

THE GIZA PARK PROTOCOL HOW TO START THE SUN

#15) The Geyser Sphinx is a static, ground-based cloud seeding generator system. This design is still in use today to dispense seeding material into the clouds. The Sphinx has an 18-inch-diameter hole in her head, which was sealed in the early 1900s to prevent people from climbing inside the tunnel system it connects to.

Note: The maximum pressures produced by the Cadman Pump can reach 3,200 psi at the head of the Geyser Sphinx. This could send a mighty deluge of water shooting vertically into the desert sky, reaching almost 6,500 feet into the atmosphere.

This design is ideal for creating low-level clouds and seeding clouds for weather and climate modification. Even today, modern clouds—most likely Tesla-designed—can be generated from ground level (0 feet elevation, sea level) up to 6,500 feet in the sky.

See two models below:

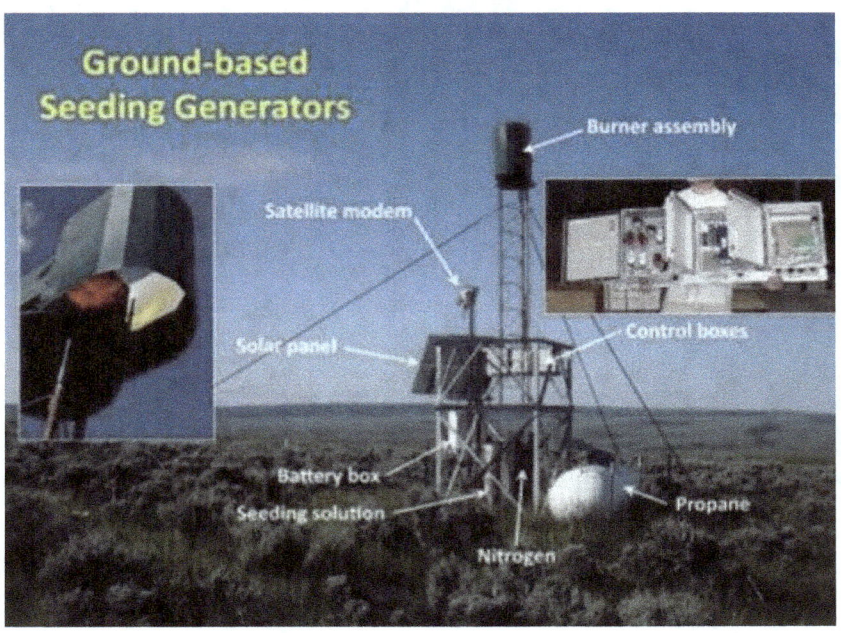

THE GIZA PARK PROTOCOL HOW TO START THE SUN

This design takes advantage of the topography, prevailing winds, and the Earth's Coriolis effect on weather patterns.

It also creates a low-pressure system that follows the indigenous weather patterns historically known as Hurricane Alley. This system travels across the hot Sahara Desert, blasting the North American continent. This system's design may have contributed to the formation of the Gulf of Mexico; additional details will follow.

#17) Abu Rawash

Abu Rawash lies on an elevated hill about five miles north of the Great Pyramid. Its role is quite different from that of other systems, as it serves a distinct function at the final stage of these structures.

Abu Rawash is positioned last and high on a hill because it provides the venting system for the industrial underground tunnels on the Giza Plateau. Just like a home has a vent pipe exiting the roof to ensure drains flow smoothly and toilets flush as designed, Abu Rawash serves a similar function. With seawater flowing through this system, salt accumulation is a significant maintenance issue. I believe this may be why all the structures at Abu Rawash are either demolished or severely damaged, appearing as if they were blown up—hence its nickname, the "Exploding Pyramid."

I have my own theories on what destroyed this ancient site, which I will discuss later.

Nile River

It is incredible how perfectly this entire system was engineered, designed, and functioned.

This system is so precise that only two structures—those connected to the Great Pyramid's Causeway and Abu Rawash's Causeway—have unused water that ultimately flows into the Nile River.

THE GIZA PARK PROTOCOL HOW TO START THE SUN

A civilisation with superior knowledge and skills must have designed this ingenious system. They even ensured that the water discarded into the Nile was ecologically managed so that it did not disrupt the river's delicate ecosystem. Perfect.

#16) Wall of the Crow

The Wall of the Crow remains one of the greatest mysteries of the Giza Plateau. Most researchers and archaeologists believe this wall was built to separate the pyramid systems from the Workers' Village, where they claim labourers lived during construction.

I strongly disagree and can't help but laugh at the idea. I even spoke with Mark Lehner, the archaeologist contracted by Edgar Cayce's Association for Research and Enlightenment. Mark dedicated painstaking years to this project and did a superb job mapping out the Giza Plateau. His excavations extended beyond the Workers' Village on the south side of the wall to include the north side as well. His findings provided valuable data that helped me realise the multiple functions of this wall—most notably, I believe it was designed to prevent water from flooding the village.

At the time, Mark was convinced my theories were incorrect. To this day, he insists it's just a stone wall.

Let me describe this structure now, and I will present its true purpose at the appropriate time.

The Wall of the Crow is built from massive, megalithic stones stacked atop one another. It is 650 feet long and was constructed just southeast of the Electric Pyramid's Causeway. It runs almost parallel to the causeway but has a slight angle when viewed from the west.

The WOC is a colossal, heavily built structure, measuring 33 feet tall and 33 feet wide. (Note the numerological significance of 33, and that 650 reduces to 11.) The wall features a large hole at ground level, resembling a short tunnel.

THE GIZA PARK PROTOCOL HOW TO START THE SUN

The lintel stone supporting the top of this hole is monolithic and is the largest stone used in any of the structures—including the Great Pyramid itself. For comparison, the Great Pyramid contains five high-crystal-content granite stones stacked one atop the other in the ceiling above the Reaction Chamber, each weighing 70 tons. The lintel stone over the hole in the Wall of the Crow is even heavier.

Additionally, on the north side of the excavation site, closest to the pyramids, a berm was backfilled with an interesting non-conductive material. This berm stood 33 feet tall—matching the height of the wall—and extended 90 feet in the direction of the pyramids, spanning the entire 650-foot length of the wall.

Why? Why? Why?

(See image.)

THE GIZA PARK PROTOCOL HOW TO START THE SUN

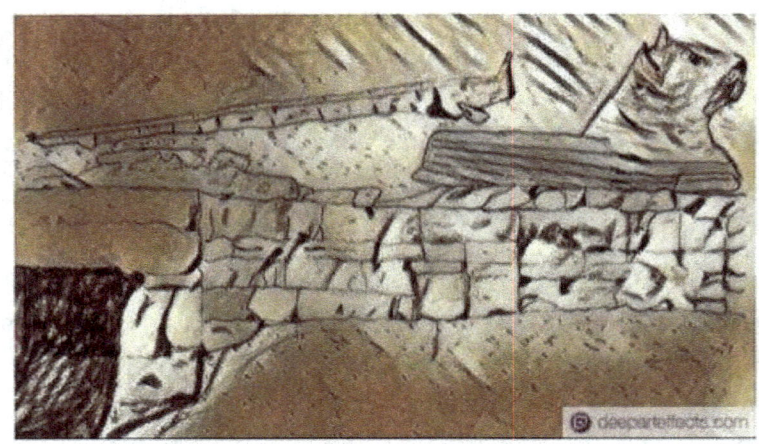

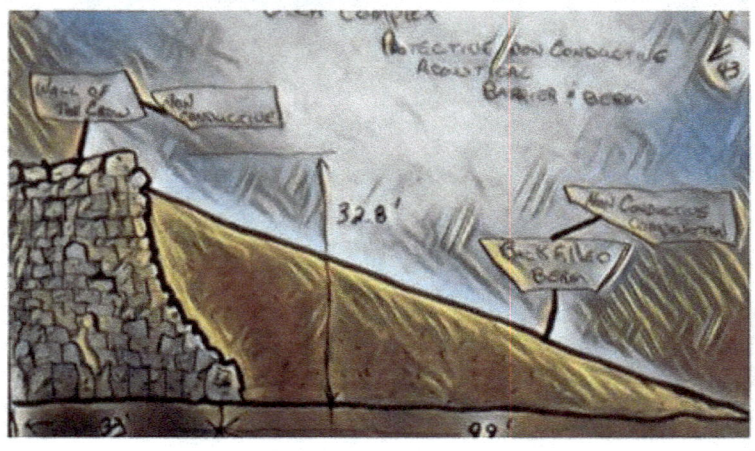

Chapter Twelve:
The Bosnia Pyramids Pleiades Correlation Theory

The Pleiades Astro-Blueprint for the Bosnian Pyramids

Following my presentation at the 7th International Scientific Conference on Bosnian Pyramids, I gave a lecture in which I introduced the community to the first perfect example of why the bull was sacred in Egypt.

Detailing the Egypt-Taurus Correlation Theory led to a conversation between Dr Semir Osmanagić and me about performing an alignment of the Bosnian Pyramid system. He suggested that I focus first on the Pleiades, providing great insight.

I immediately delved into the research, and below are the remarkable results that offer absolute corroboration of the star alignments and other correlations with the Egyptian Pyramid system.

I see this as an incredible opportunity to present these unbelievable, profound discoveries, which will ultimately change not only Bosnia's history but also the history of humanity forever.

I will describe the Bosnia-Pleiades Correlation Theory using the same basic format as the Egypt-Taurus Correlation Theory. I will incorporate images and the same flowchart method I used previously. This structured approach simplifies complex systems, making the concepts accessible to everyone.

See images: Let's begin at the location below, with each site's corresponding Pleiades star names and descriptions.

THE GIZA PARK PROTOCOL HOW TO START THE SUN

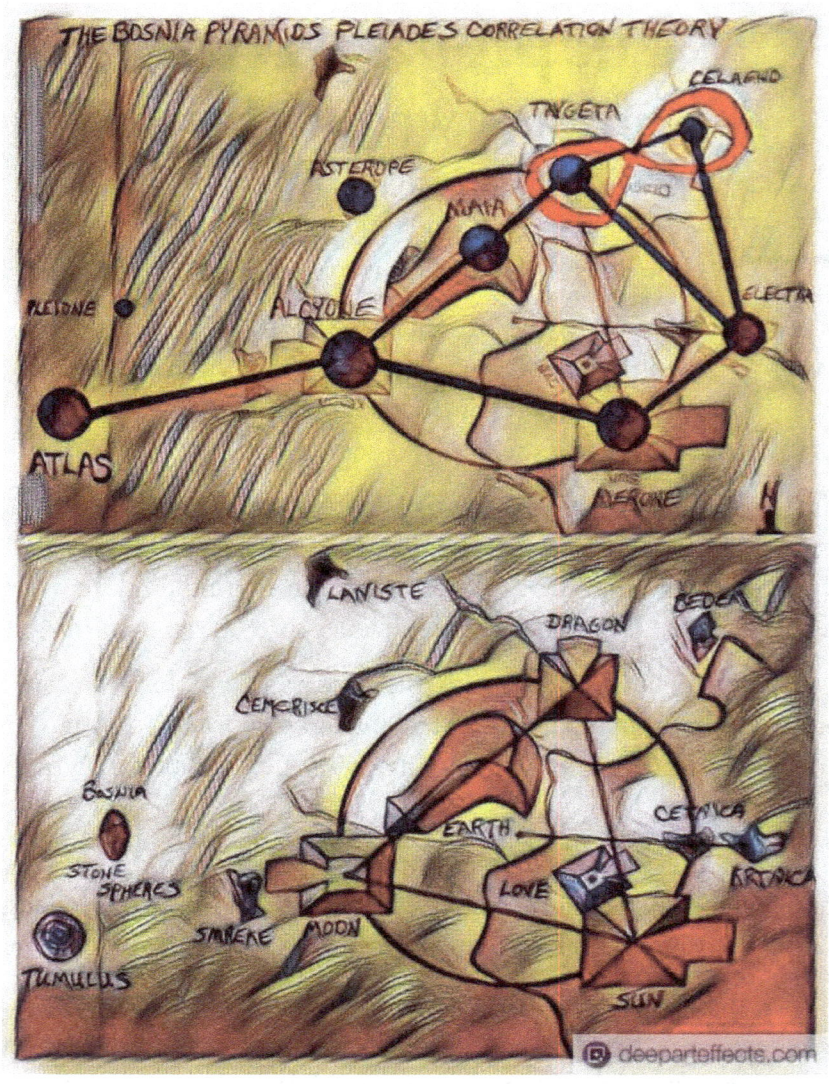

THE GIZA PARK PROTOCOL HOW TO START THE SUN

#1) **TUMULUS – Masculine Energy**

Pleiades star, **ATLAS – Masculine Energy**. Atlas was a Titan in Greek mythology. Zeus condemned him to hold up the heavens. Hesiod, the great Greek poet, said Atlas stood at the ends of the Earth.

At the Atlas Mountains, the Atlantic Ocean was once called the Sea of Atlas.

In this first identification from my panoramic perspective, I want to address Atlas as the only masculine energy associated with the Pleiades constellation. His presence is undeniable—he holds the seven sisters and their mother on his shoulders, supporting the heavens so their brilliance may shine upon humanity and the Earth.

THE GIZA PARK PROTOCOL HOW TO START THE SUN

THE GIZA PARK PROTOCOL HOW TO START THE SUN

THE GIZA PARK PROTOCOL HOW TO START THE SUN

THE GIZA PARK PROTOCOL HOW TO START THE SUN

Image: Let's begin at the geographical location below, with the Pleiades star names and descriptions to follow.

#2) **Bosnia Stone Spheres – Feminine**

It is most appropriate, then, that the Bosnian Stone Spheres would be the closest to Atlas.

Is this merely coincidence?

Pleiades star, PLEIONE – Feminine energy

Not only does Atlas hold up the Earth, but Pleione is also the mother of all the Pleiades sisters. Her name alone means "to increase in number," and she is responsible for multiplying flocks of birds.

Image: Let's begin at the geographical location below, with the Pleiades star names and descriptions to follow.

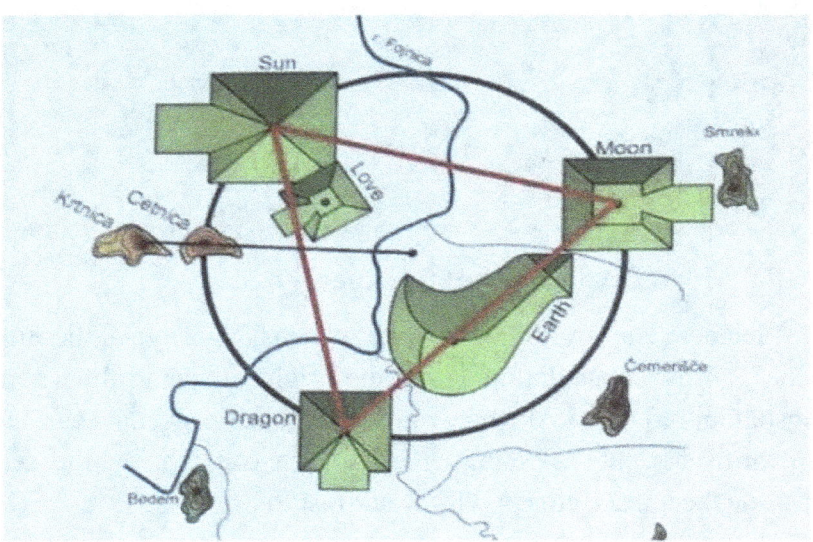

The image above is a general diagram of the Layout of the Bosnian Pyramid Systems.

You can easily see their locations and their relationship with each other.

THE GIZA PARK PROTOCOL HOW TO START THE SUN

To me, it looks like another ancient industrial-designed system.

In this next image, you can see the expected shape and design of several pyramids about one another to gain a perspective.

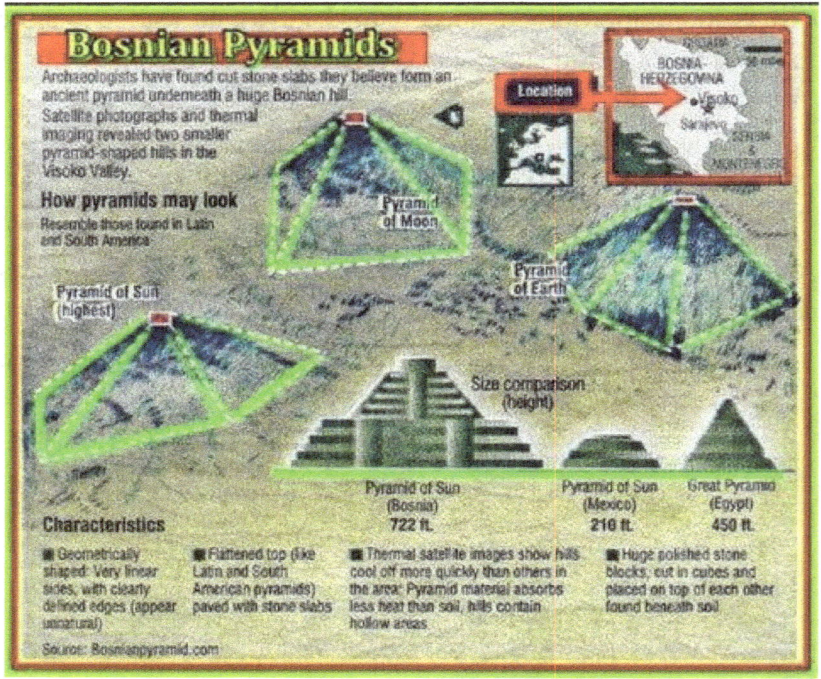

#3) Moon Pyramid-Feminine energy

Pleiades star ALCYONE, also called the "Right eye of the Bull" in the Taurus Constellation. Also, the brightest star in the Taurus Constellation. Due to Alcyone's grief over a sailor lost at sea, Zeus transforms her into a "Kingfisher" so she can live on the sea's coastline, keeping a close vigil for her lost love.

Her name causes love, clarifies intuition, increases manifestation potential, and trusts yourself.

You can see the mountain-like construction of this Earthing Pyramid in the image above.

THE GIZA PARK PROTOCOL HOW TO START THE SUN

#4) Mother Earth Pyramid- feminine energy,

Pleiades star, MAIA, the Great Mother, The Nurse of the Earth, Gaia, Earth Mother, and the mother of the Greek God Hermes Trismegistus.

She is the oldest star and sister.

#5) Cemensce-Pleiades star, Feminine, ASTEROPE, her name means "Lightning."

As you read this text, you will understand the significance of lightning in Bosnia and learn much about its nature and characteristics. Take it from me, Lightning Leo, Lol.

THE GIZA PARK PROTOCOL HOW TO START THE SUN

I will give a systems analogy at the end of this presentation.

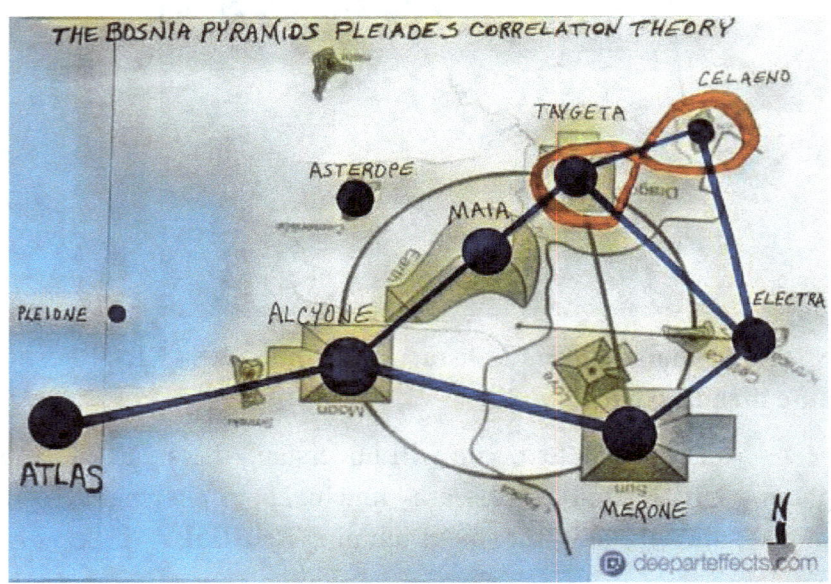

#6 & #7) #6) Dragon Pyramid #7) Bedem and Fojnica River, Branch.

The red Figure 8 of Pl shows the Feminine, TAYGETTA, and Double star clusters orbiting each other and becoming close and

THE GIZA PARK PROTOCOL HOW TO START THE SUN

separate. This indicates that only three stars exist when their orbits coincide at specific orbit cycles.

Taygetta was a Feminine Heroine who birthed Sparta with Zeus.

An interesting circumstance is that the Fojnicka River is a crucial component with two stars connected with its northern and southern branches. I wonder if the apparent fact that the river floods annually following a season of heavy snowfall might be a factor.

This image below is an excellent way to express Celeaneo's point that the other sisters' brilliance outshines the Dark One.

#8) Cetnica-feminine energy,

Pleiades star CELAENO, whose name means "The Dark One," is often not visible to the naked eye because she is the middle star between two bigger and brighter stars.

THE GIZA PARK PROTOCOL HOW TO START THE SUN

Below is an Artist's rendering of what the Bosnian Pyramid of the Sun looks like beneath its 1,000 years of foliage growth, which camouflages the structure's true likeness.

The image below compares the Bosnian Pyramid of the Sun to the Egyptian Great Pyramid to demonstrate its enormousness.

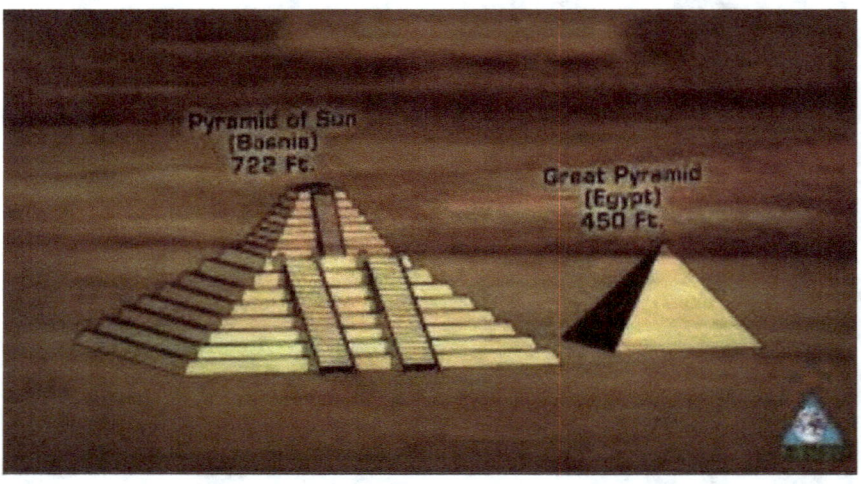

Indeed, with the data available today, the Bosnian Pyramid of the Sun is equitably the most prominent artificial structure on Planet Earth until another one larger is discovered.

#9) Pyramid of the Sun-feminine energy,

Pleiades star ELECTRA; her name means "The Bright One" or Amber, radiance, incandescent, shining, brightness.

THE GIZA PARK PROTOCOL HOW TO START THE SUN

Electra's greed for power may have influenced Cleopatra herself, who murdered her sister for the throne, while Electra murdered her mother for the same reasons.

#10) The Fojnica River, Branch

Pleiades star, MEROPE, The Missing or disappearing star.

I am including the information below because it is critical collaborative evidence that needs sharing.

#11) The Sun, Summer Solstice, Helical Rise of Sirius, 7,000 BC @ 63 degrees, Tropic of Cancer.

#12) Sirius, Summer Solstice, Helical Rise of Sirius, 7,000 BC @ 114 degrees, Tropic of Cancer.

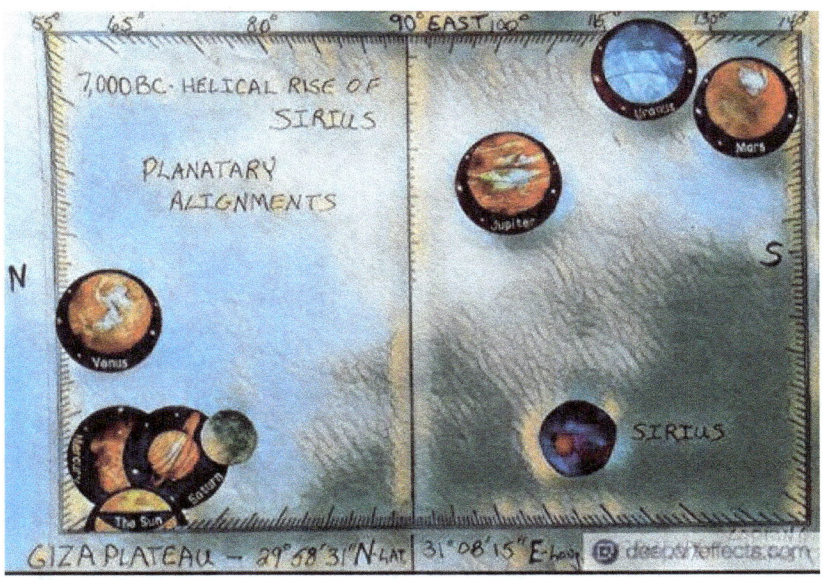

This is the same image I displayed when I detailed the Helical Rise of Sirius in Egypt. I did so because they are nearly identical in alignment because of their geographical proximity. The Farthest North the Sun Travels North is to the Tropic of Cancer, which aligns 63 degrees East from Egypt and Bosnia.

THE GIZA PARK PROTOCOL HOW TO START THE SUN

See the image of the planetary alignments below of Bosnia and compare it to those of Egypt.

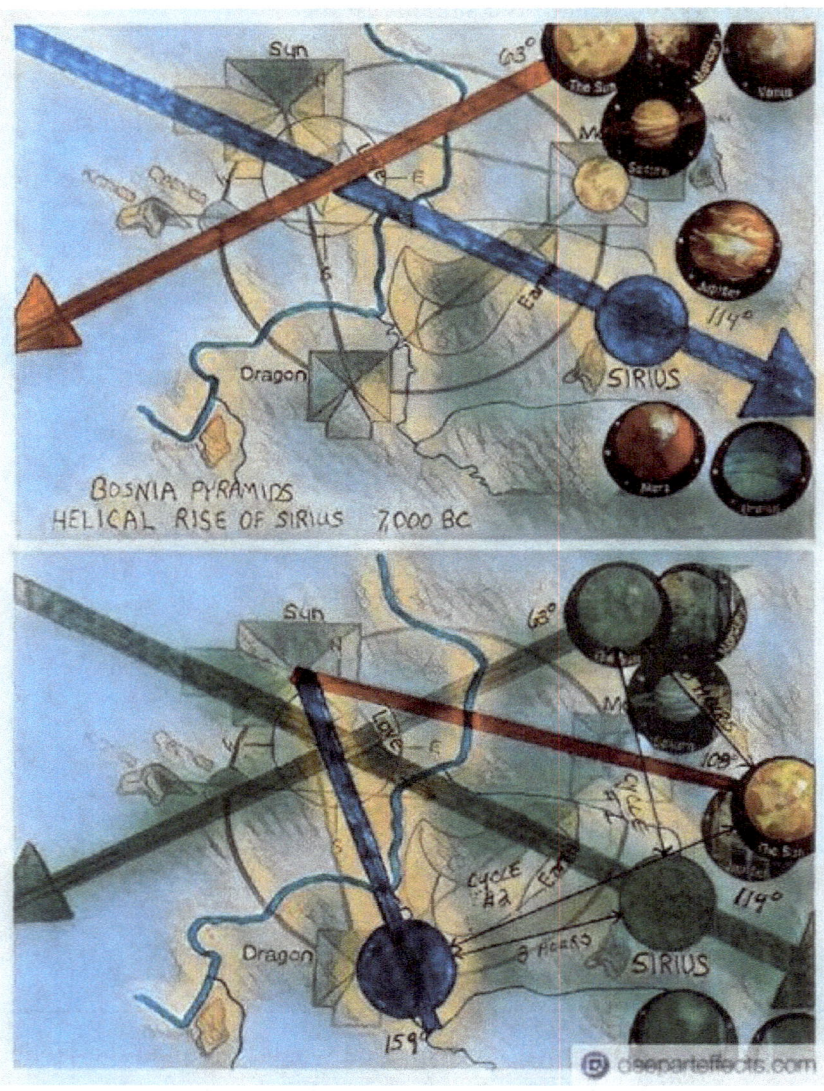

THE GIZA PARK PROTOCOL HOW TO START THE SUN

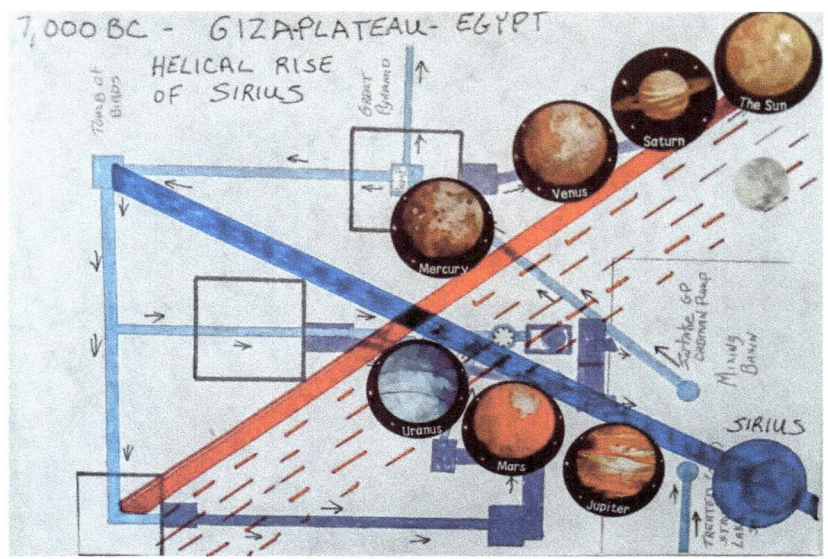

As above, so below.

Also, I communicated with Dr. Semir Osmanagich, who has done exemplary work on the correlation of geography with sacred geometry at Visoko and the Bosnian Pyramid Systems.

See the images below to confirm:

THE GIZA PARK PROTOCOL HOW TO START THE SUN

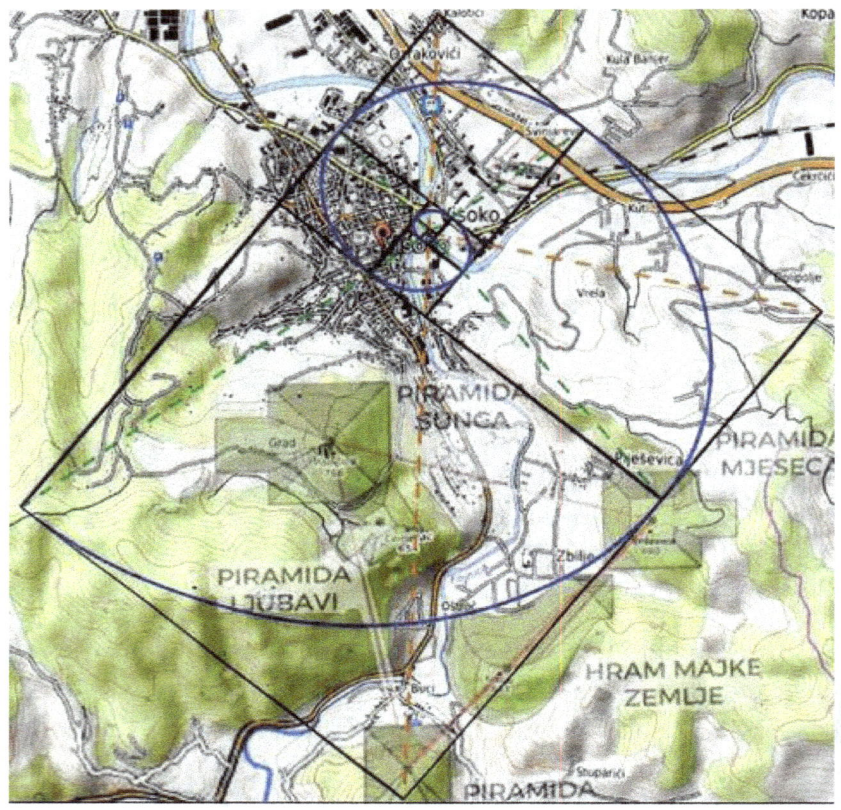

It is easy to see the use of the Fibonacci spiral and numbers in series and sequences, close to representing the golden spiral geometrically and symbolically. The golden spiral is represented in nature in pinecones, snail shells, and pineapples.

This first image shows how the city of Visoko has been designed and relates to its close neighbor, the Bosnian Pyramid Systems.

Looking at the Fibonacci spiral connects to the mouth of the Fojnica River and River Bosnia, the top of the Bosnian Pyramid of the Moon, the Temple of Mother Earth curve, and the top of the Cetnica Pyramid Hill.

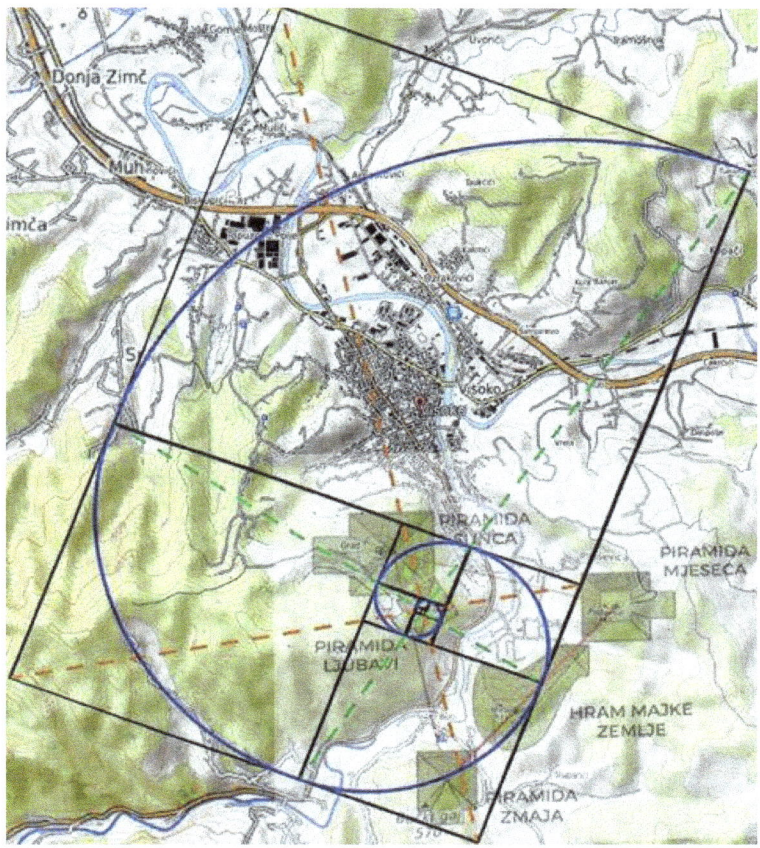

This second image shows how the Fibonacci Spiral effortlessly fits equally with the Bosnian Pyramid Systems and relates perfectly to Visoko.

The Fibonacci (golden section) spiral above is the top of the Bosnian Pyramid of Love, the top of the Pyramid of the Sun, the top of the Temple of Mother Earth, the top of the Pyramid of the Dragon, and the top of the tumulus in Vratnica.

I have provided physical evidence from many different perspectives, attempting to remove doubts that the Bosnian and Egyptian Pyramids are not sacred geometries designed and astronomically aligned.

THE GIZA PARK PROTOCOL HOW TO START THE SUN

This last image is called the "Perfect Circle" by Mathematicians, and I offer it as another "Perfect" example of the Intelligent Design of Bosnia and Egypt.

I cannot imagine any scientist or academic who can review all the data, including a physical inspection of the Geography and structures in Bosnia, and claim this region is natural and not intelligently designed.

It is evident that the Science community is CORRUPT to the Core $$$$$ and has lost all its Integrity.

THE GIZA PARK PROTOCOL HOW TO START THE SUN

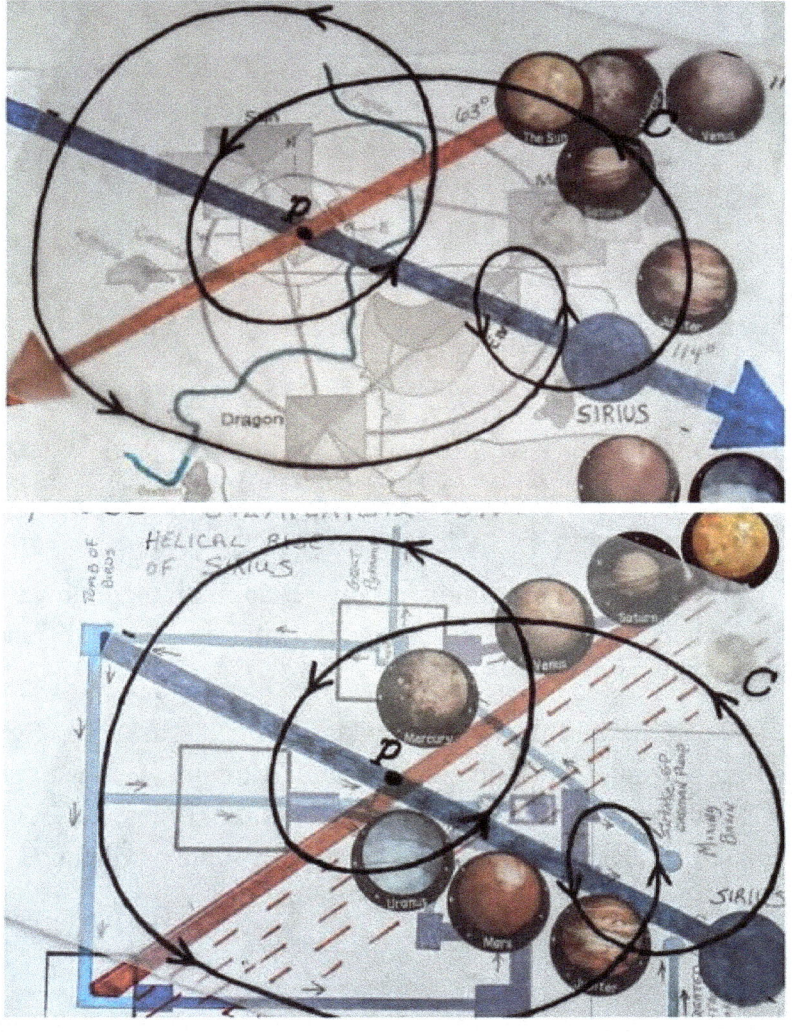

Before I end this discussion, I want to elaborate further on the perfect astronomical alignments and synchronized Earth structures and natural systems in the Northern Hemisphere on the Summer Solstice each year.

THE GIZA PARK PROTOCOL HOW TO START THE SUN

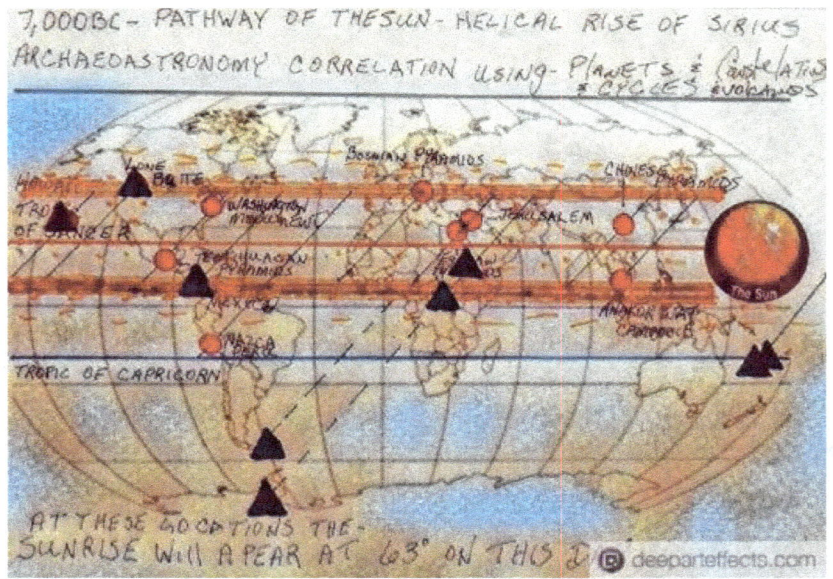

This image I created following my research on the Pathway of the Sun when the Sun is aligned perfectly with the Tropic of Cancer on Summer Solstice.

The Chinese Pyramids, Angkor Wat, Jerusalem, Egypt Pyramids, Bosnian Pyramids, Washington Monument, Washington D.C., Teotihuacan Pyramids Mexico, Peru Nazca Lines, Lone Butte Subglacial Volcano, Hawaii, and many mighty volcanos all perfectly aligned with this same sky on Summer Solstice.

See individual charts on the collage below:

THE GIZA PARK PROTOCOL HOW TO START THE SUN

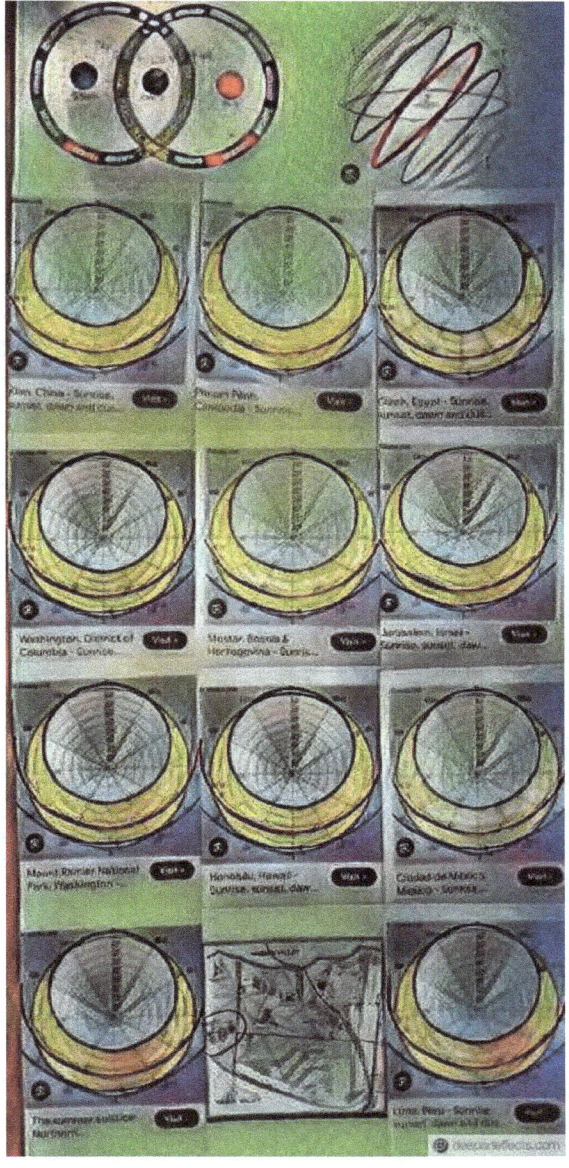

All these locations verified have precisely the same 63-degree relationship and cosmic alignments that day.

Chapter Thirteen:
The Teotihuacan Aries Correlation Theory

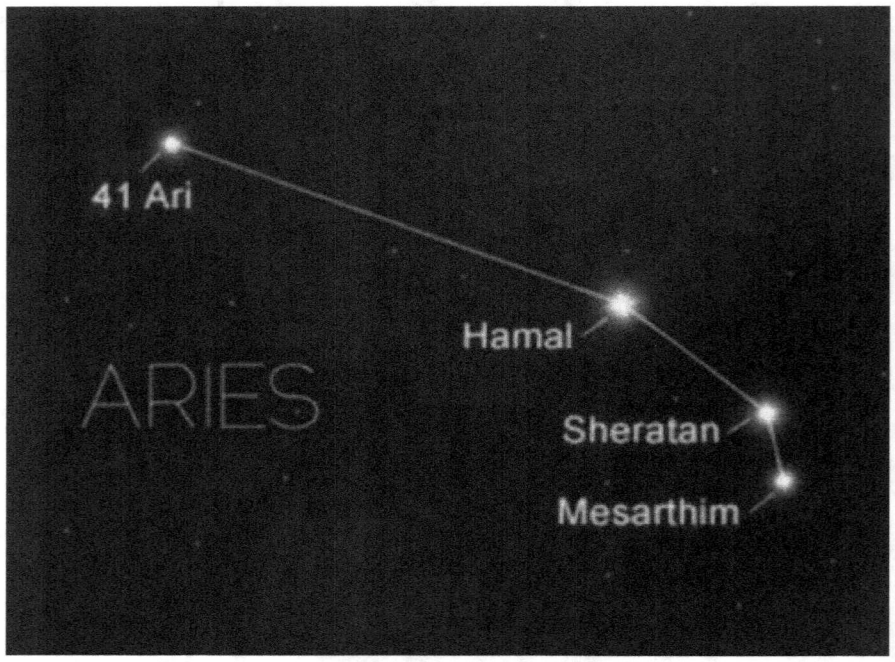

Following my research on the Egypt-Taurus and Bosnia-Pleiades Correlation Theories, my confidence soared, and I decided to explore just one more system—this time in Mexico.

This journey of discovery has been truly mind-blowing, as a significant amount of new information has been gathered regarding the construction, operations, and reasoning behind these ancient systems.

I have already identified several new approaches to applying the Astro-Archaeology Alignment method.

THE GIZA PARK PROTOCOL HOW TO START THE SUN

The greatest revelation so far is the recognition that the stars in constellations serve as system locators. Natural systems such as rivers, river junctions, volcanoes, hydraulic dives, cenotes, tectonic plates, keylines, vortex systems, large mineral and metal deposits, and seas are all included in these alignments.

Now, let us examine the Aries constellation and compare it to the systems at Teotihuacan.

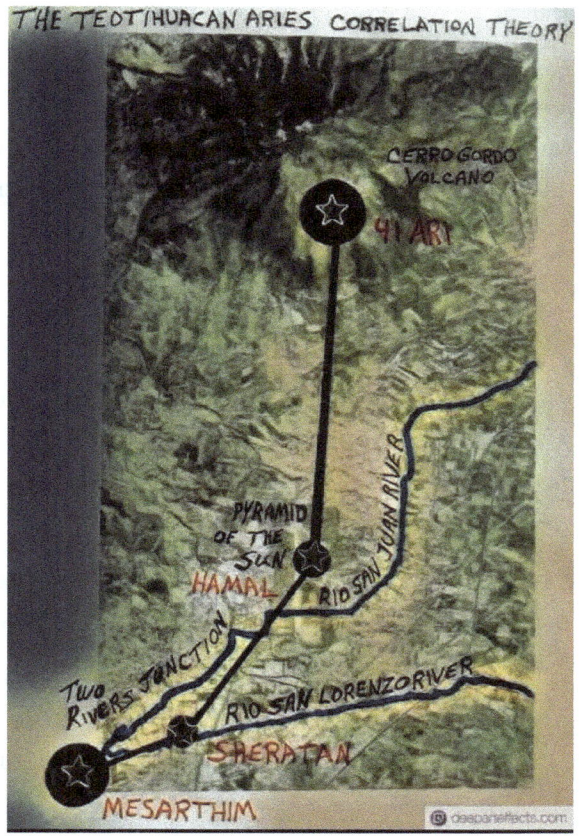

Typically, Aries is visible in northern latitudes as just four naked-eye stars. One thing I noticed the first time Heidi and I visited the Teotihuacan Pyramids—one of my favourite pyramid sites in the world—was a striking alignment that immediately caught my attention.

THE GIZA PARK PROTOCOL HOW TO START THE SUN

Wherever I conduct research and investigations, I always look for symbolic messages. I see them everywhere—the Geyser Sphinx in Egypt is a prime example.

In Bosnia, I observed that the design and arrangement of the systems carry symbolic meaning. Additionally, I have found overwhelming evidence suggesting a strong connection between the pyramid systems in Bosnia and Egypt. These two independent systems appear to have been designed to work together towards a common goal: **to end the Ice Age.**

In Mexico, my first major observation was how the Pyramid of the Moon was deliberately shaped to resemble Cerro Gordo, the volcano to its north. Once you recognise this, the resemblance is undeniable.

Let's begin.

1) Cerro Gordo Volcano

The Cerro Gordo stratovolcano had a violent birth. Its first eruption was hydromagmatic in nature, giving rise to a large maar—a lagoon that forms in the crater of an ancient volcano.

A hydromagmatic eruption is classified as a Surtseyan eruption, which is a type of phreatomagmatic eruption. These eruptions are highly explosive due to the intense interaction between rising magma and water from a lake or the sea.

Radiocarbon dating of organic material within its magma dome cap indicates that Cerro Gordo's last known eruption occurred around **3,700 BC**.

2) Aries Constellation Connection

The Aries constellation star **41 ARI** is located **166 light-years** from our Sun. Its surface temperature is approximately **12,000 Kelvin (21,140°F)**—blazing hot!

THE GIZA PARK PROTOCOL HOW TO START THE SUN

This raises an intriguing question: **How does this happen?** Who decided, "Let's match up scorching stars with volcanoes, pyramids, and rivers?"

The alignment appears to be a **perfect match** between celestial and terrestrial features.

When you examine Cerro Gordo's position relative to the Teotihuacan complex—just three miles away—it becomes difficult to dismiss the possibility of a deliberate physical relationship.

Chapter Fourteen:
Hieroglyph of Ani, Plate # 2 "Sunrise"

Theory

As I deepened my understanding of multiple reverse engineering projects, key components of the Egyptian weather and climate modification systems became more refined. Alongside my systems analysis, I balanced numerous other investigative research projects across the globe.

During this educational journey, my research into the actual influence of the **Summer Solstice period**—specifically its connection to the **Heliacal Rise of Sirius** over the Giza Plateau—led me to a particularly significant hieroglyph.

Let us first identify the location where this artwork appears. I will present substantial physical evidence, including findings from **Mr. Zahi A. Hawass**, Secretary General of the Supreme Council of Antiquities in Egypt, based on his extensive investigations into *The Discovery of the Osiris Shaft at Giza*.

The image and measurements provided by **Mr. Hawass**—along with my own desk-based analyses—were instrumental in determining the **dynamics of water flow** in this ancient system.

THE GIZA PARK PROTOCOL HOW TO START THE SUN

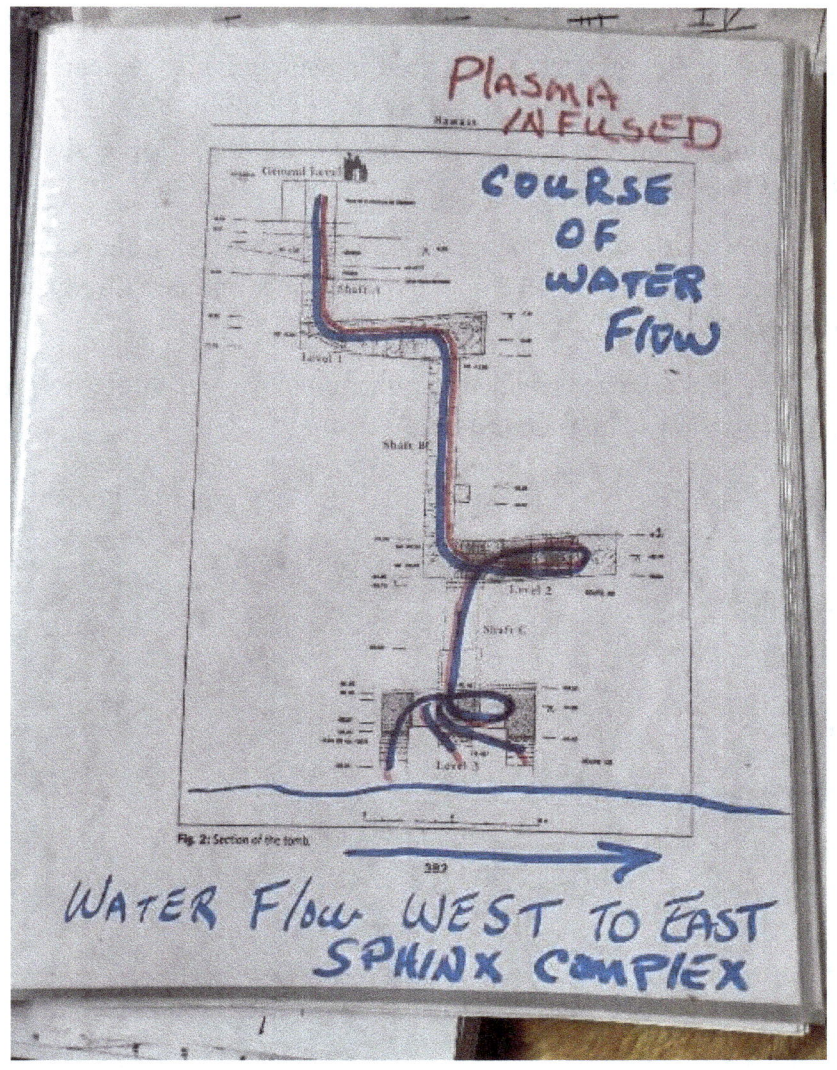

THE GIZA PARK PROTOCOL HOW TO START THE SUN

This research project was a major discovery for **Mr. Zahi Hawass**. As he reflected on his accomplishments, he dedicated his findings to his mentor, **Dr. David O'Connor**, the esteemed Egyptologist under whom he trained at the **University of Pennsylvania**.

I cannot think of a better example of physical, collaborative evidence, especially when combined with other research discoveries from credible sources.

Mr. Hawass has provided the image below, which presents an **overhead view of the second-level layout**.

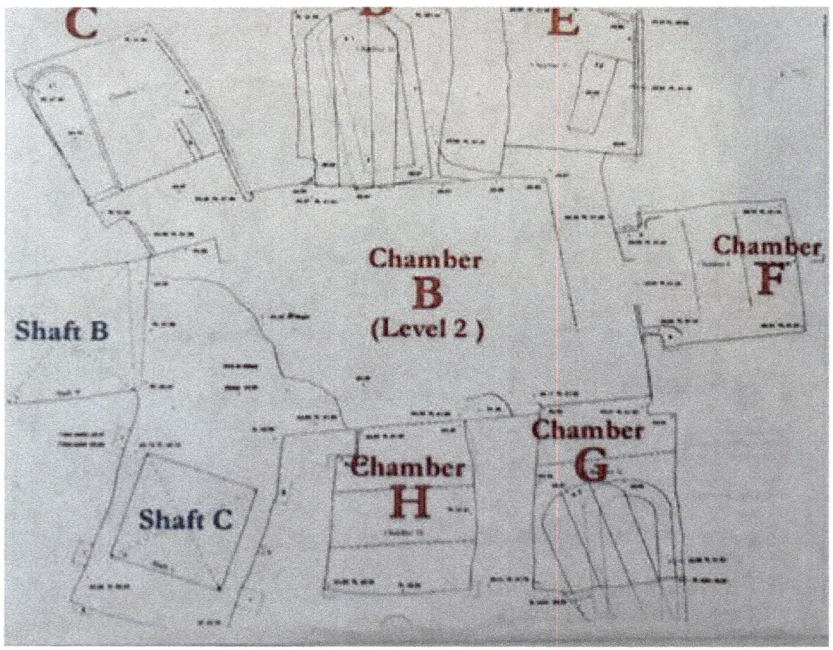

Now that we are familiar with the **second-level** system design, let us shift our focus to the **third-level systems** below.

To begin this discussion, I want to highlight one of the most famous **Egyptian artifacts** housed in the **British Museum**—the hieroglyph *"Sunrise"*, featured in the **Papyrus of Ani**.

THE GIZA PARK PROTOCOL HOW TO START THE SUN

Pay close attention to the layout of this image. Notice that the **outer structure** of the papyrus has a **rectangular shape**, much like a picture frame. Traditionally, such frames were **hand-drawn** by highly skilled artists of their time. In this case, the artwork was created by an **Egyptian scribe named Ani**.

THE GIZA PARK PROTOCOL HOW TO START THE SUN

It is easy to recognise that the **sides** of the image appear **rough and jagged**, resembling **stone**.

The **bottom side**, however, looks **flat and level**, carefully crafted with **three drains**—a feature strikingly similar to **Osiris's Tomb**.

As for the **top**, it has a **slightly curved radius**—a detail we will revisit shortly.

This **rectangular outer structure** strongly resembles **Osiris's Tomb**, as shown in the image above.

THE GIZA PARK PROTOCOL HOW TO START THE SUN

By comparing the **Papyrus of Ani** with the actual **Osiris Tomb drawings**—provided by **Mr. Zahi Hawass** and located precisely **108 feet below ground**—we must seriously consider the possibility that this scene represents events within the Osiris Tomb itself.

Following this theme, as we carefully examine each piece of evidence in detail, I am eager to reveal my **cosmic discoveries**.

The image below is a drawing provided by **Mr. Hawass** based on his research into the Osiris Tomb and its measurements. I have chosen to **colour-code** and identify the **masculine (+) positive red** and **feminine (-) negative green** energies that I sensed were present.

I plan to reference this image in my upcoming analysis, where I will detail the **functions of various systems**. This treatise has been created to **identify, document, and raise awareness** of these profound connections.

THE GIZA PARK PROTOCOL HOW TO START THE SUN

Now is the appropriate time to introduce everyone to the inventory list of the papyrus of Ani, Plate #2, including the eleven systems with their component identification.

Inventory of Components

Note:

Future comparisons with **astronomical alignments** must consider that hieroglyphs and **ancient structures built at ground level** are sometimes portrayed as **mirrored images** of the celestial patterns above them.

So far, after analysing three astronomical alignments, it appears that **positive (+) masculine systems** are often represented as **mirrored images**, while **negative (-) feminine energies** maintain a **direct relationship** with their celestial counterparts. As more people contribute to this research, we may eventually identify **every system on the planet** and determine its function.

Key Observations & Systems Analysis

1) **Osiris Tomb Intake** – Functions as a **plasma water intake**, receiving energy from the **Plasma Reflux Pyramid Causeway**.
2) **Osiris Tomb – Second Level** – A site of **dynamic plasma water structuring and electrolysis**. This level is particularly fascinating, as it offers critical insights into how the **ancient Egyptians understood fluid dynamics, energy transfer, and their connection to astronomical alignments**.
3) **Osiris Tomb – Third Level** – This level is where **plasma water dissociation** and **alternating scalar waves** occur. It provides crucial clues about the **ancient Egyptian knowledge of high-energy physics and cosmic interactions**.
4) **KA** – Represented as **two arms and hands reaching outward**, this Egyptian hieroglyph symbolises **the connection between humans and their creative life force, linked to the gods**.

THE GIZA PARK PROTOCOL HOW TO START THE SUN

5) **Ionosphere** – The **F-region** of this **plasma ocean** contains highly dynamic **density structures** with varying spatial scales. These large-scale formations, which can extend **tens to hundreds of kilometres**, fluctuate based on the **time of day, season, solar cycle, geomagnetic activity, and solar wind conditions**. This **life-producing plasma** surrounds the Earth, shielded by the **magnetosphere**.
6) **Isis** – Queen and wife of **Osiris**, revered as the **goddess of love, healing, fertility, magic, and Sirius**. Isis remained the most worshipped deity in **Egyptian culture**, even under **Greek rule**.
7) **Three Monkeys** – Identified as **A-B-C**, these figures are depicted with **arms and hands raised in the KA or worship stance, directing their energy toward point #10**.
8) **Three Mirrored Monkeys** – Also identified as **A-B-C**, these mirrored figures mimic the stance of the three monkeys above, **with their energy also directed toward point #10**.
9) **Nephthys** – Sister of **Isis** and goddess of the **air**, known as the **"Mistress of the House"** and **"Protector of Osiris"**. She was married to **Set**, who murdered Osiris, **dismembered his body into 14 fragments**, and scattered them throughout Egypt.
10) **Bow Shock or Wake of the Sun** – Represents the **telluric scalar current** and the **phase shift frequencies accompanying sunrise** each day. In astrophysics, a **bow shock** occurs when the **magnetosphere of the Sun interacts with the nearby solar wind plasma**. This **phase shift** slows the **supersonic solar wind** to **subsonic speeds**, bringing it below the **sound barrier**.
11) **Super Plasma** – A **high-energy EMF (electromagnetic field) and piezoelectric-infused structured water medium** with an extensive **distribution system**. This includes **three drain systems from Osiris's Tomb**,

THE GIZA PARK PROTOCOL HOW TO START THE SUN

directing **plasma water eastward through a tunnel beneath the Geyser Sphinx**.

12) By reviewing the inventory list of players in *The Art of Ani*, we can now add Isis as collaborative evidence. She was the wife of Osiris and the most famous and worshipped Queen of Egypt in history.
13) The second piece of collaborative evidence is the inclusion of Nephthys in the play. She was Isis's sister, married to Osiris's brother, and the goddess responsible for protecting Osiris in the afterlife.
14) Please refer to the beautiful images created by Mr. Hawass for comparison.
15) I have now presented three crucial clues. Have I provided sufficient collaborative evidence to support my theories and at least suggest that the location of the play is Osiris's Tomb?
16) Let's now introduce the third level, which reveals the mechanics of transporting water to the fourth-level tunnel.
17) As we move forward in our investigations, let us focus on an extraordinary period in Egypt's history, known as the *Heliacal Rise of Sirius*. This term refers to the simultaneous sunrise of our Sun and the mythical second Sun, Sirius, every Summer Solstice. This spectacular event occurs directly in front of the Geyser Sphinx and provides further evidence as to why the Giza Plateau was chosen as the location for these immense structures.
18) This event traditionally marked the start of the rainy season when the Nile River flooded its banks, covering the infertile desert sands with rich, fertile soil, which was then used for agriculture.
19) My theories explore this in detail as we continue gathering information about this ancient and complex system.
20) Let us take a moment to look beyond our physical location at the Giza Plateau and admire the Great Pyramid. Standing

THE GIZA PARK PROTOCOL HOW TO START THE SUN

in front of the Geyser Sphinx, one must wonder—what exactly is the Lioness Sphinx looking at?

21) Once I determined the timing of the "Great Flooding" in this region, which lasted from around 8,000 BC (the Age of Cancer) to 3,000 BC (the Age of Taurus), I decided to investigate further. I narrowed my focus to cosmic alignments during these cycles, covering 5,000 years of extreme rains and flooding.

22) I selected a date 1,000 years into this rainy period—7,000 BC—around the Summer Solstice and the Lion's Gate Portal period. This corresponds with the event historically known as the *Heliacal Rise of Sirius*, marking the annual beginning of the rainy season and the period of the *Great Floods*.

23) My chosen location was the precise site of the Geyser Sphinx, with GPS coordinates of 29.58'31" N and 31.08'15" E—looking eastward towards the Pleiades, or what is today known as the Octagon.

24) My reasoning is unlike anything previously proposed. My research has been extensive, and my pursuit of identifying quantum energy streams is unconventional and outside mainstream science.

25) The conclusions on which I base my theories have consistently demonstrated that some of the pyramid systems around the globe were designed to serve a specific function.

26) If we truly wish to understand this planet, we must adopt a new perspective—one that differs from anything previously considered.

27) I wanted to determine the positions of the planets in our solar system during the *Heliacal Rise of Sirius*, as I had already concluded that this was the moment when the Weather and Climate Modification System at the Giza Industrial Park would have been activated.

THE GIZA PARK PROTOCOL HOW TO START THE SUN

28) Now, imagine yourself standing between the paws of the Geyser Sphinx, facing east.
29) You will not believe your eyes.
30) Below is a sky diagram of cosmic planetary alignments, which I have detailed using various methods to identify their significant energy patterns.

The Greatest Discovery on the Giza Plateau, Osiris's Tomb

When I began my forensic research project on the Great Pyramid, I had no genuine interest in pyramids beyond general curiosity. I had long believed they were all tombs, with one belonging to the Ancient Egyptian Pharaoh Khufu.

Over time, I have learned to recognise the art of fabricating history for economic gain. Nowhere on the planet serves as a better example of this than Egypt.

As I delved deeper, I realised that much of the information I had learned and accepted as truth was corrupted—especially regarding timelines and dates. It became apparent that I had been researching a textbook fantasy, shaped by various imaginations to serve an agenda.

THE GIZA PARK PROTOCOL HOW TO START THE SUN

This agenda originated with the early migration of the Egyptians, who crafted an incredible tale—one they documented—beginning with the Great Exodus to Israel.

Eventually, I systematically examined and reverse-engineered every structure on and beneath the Giza Plateau. Early on, I recognised that the location and alignments of most structures were energetically perfect—except for three.

The one I want to focus on now is the so-called **Tomb of Osiris**.

Why is this oddly designed and misaligned structure so different from the three pyramids and the Sphinx?

Why was it constructed in such a complex manner, with three levels descending straight into the earth—below the Great Pyramid's foundation—down to a depth of **108 sacred feet**?

Another curiosity: 108 is a sacred number. Was it truly a tomb? An ancient religious site, perhaps?

If not a tomb, then what?

This project was demanding in every way. There were so many unanswered questions, and standard logical explanations were lacking, making it difficult to form a hypothesis.

Paradoxically, the challenge of finding logical industry methods allowed my creativity to flourish without restriction.

Using the only tools I could rely on—my intuition and confidence—I was driven forward, knowing my discoveries related to plasma production had been identified. I had already determined that the Tomb of Osiris involved quantum-entangled scalar energies permeating a structured, plasma-infused water medium.

How did I reach this conclusion? It required effort and a shift in perspective—changing how I approached, observed, and identified global pyramid systems.

THE GIZA PARK PROTOCOL HOW TO START THE SUN

Using the precise measurements provided by Zahi Hawass, I worked to determine the orientation and alignment of this peculiar structure. I was already questioning its purpose.

Once I was confident in my calculations, I decided to look east—like the Sphinx—using an angle of **121.5 degrees east / 243 degrees west**.

Further research led me to use computer models to track this **121.5-degree** alignment as a compass heading. My focus shifted westward across the sands of the Egyptian Sahara Desert, adjusting to **243 degrees west**.

However, I prefer using **121.5 degrees east**, as that was the first significant number I discovered.

Now, as I present this information near the summation of my findings, your understanding of pyramid systems may have expanded—perhaps even requiring some reconsideration.

I will include images from my desk, drawing from my knowledge and experience in **flow dynamics**, **air conveying systems R&D**, and **30 years of studying tidal hydraulic flows while living afloat**.

These images illustrate how I measure and map **bio-energies of all types, including scalar energy**.

Refer to the collage image of the energy diagrams on the second level for further insight.

THE GIZA PARK PROTOCOL HOW TO START THE SUN

I hope that by sharing my methods, I may encourage others to perceive invisible energies in a new way. As Bruce Lee famously said, *"Be like water, my friend."*

The fascinating thing about **scalar energies** is that they move even faster than Bruce Lee himself.

Below, I present images of the third level—once regarded as the **Tomb of Osiris**.

THE GIZA PARK PROTOCOL HOW TO START THE SUN

I used Hawass's drawings to identify the polarity of the structures and the presence of **plasma-infused super water**.

Below are two different **artists' and researchers' interpretations** of the three levels of the **Tomb of Osiris**.

The illustration on the right highlights the **four limestone bedrock pillars**. But why were the original pillars **crushed and destroyed**?

THE GIZA PARK PROTOCOL HOW TO START THE SUN

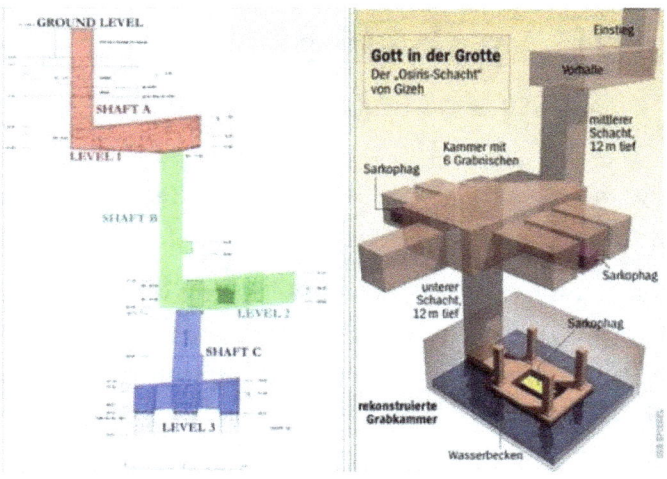

Take note of level 3 on the left image. These are the three drains from the Osiris Tomb into the 4th-level tunnel that runs directly to the Geyser Sphinx.

See the collage below:

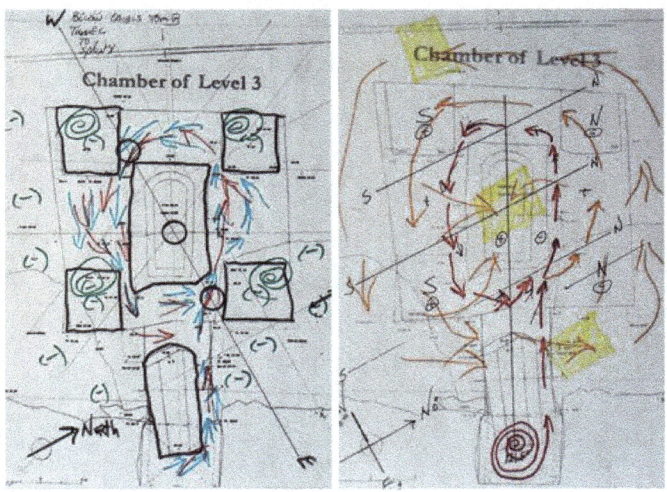

Energy Analysis of the 3rd Level of Osiris's Tomb

Below is an image of the **4th-level tunnel** and the **drainage system** from the **Tomb of Osiris**.

This **super-fertile plasma liquid** is being whisked away by the velocity and dynamic movement generated by the **3,200 psi Cadman Pump**, exiting through an **18-inch port** at the top of the **Geyser Sphinx's head**.

From this **ground-based cloud seeding system**, the water is ejected **over 6,000 feet into the lower atmosphere**.

Do I even need to suggest that **this could be how the Sahara Desert once turned green**?

It is well-documented that **plasma-infused water turbocharges plant growth**, making it clear that it could facilitate **lush vegetation in a sandy environment**—as long as **regular rainfall** occurs.

THE GIZA PARK PROTOCOL HOW TO START THE SUN

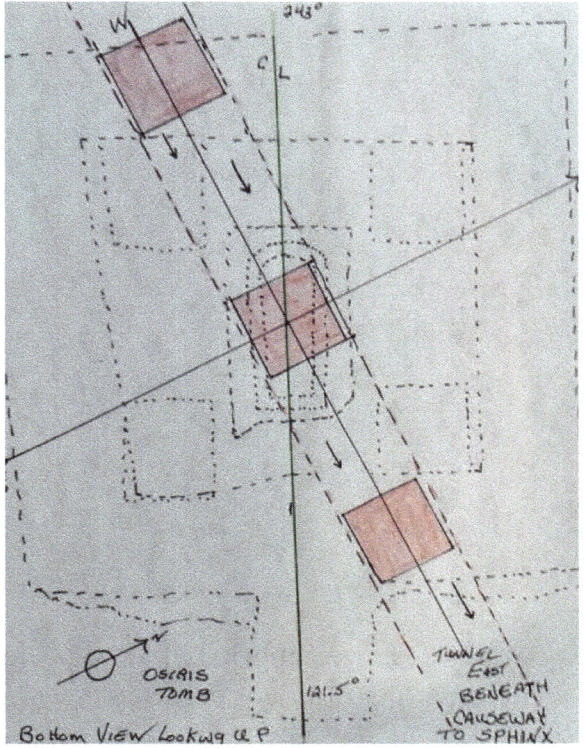

As we can see, **every detail is meticulously designed**.

This chapter is dedicated to **exploring the function of the Tomb of Osiris** and my belief that it is connected to the **Papyrus of Ani, Plate #2**.

The following visuals will illustrate my perspective and help convey what I am attempting to describe.

THE GIZA PARK PROTOCOL HOW TO START THE SUN

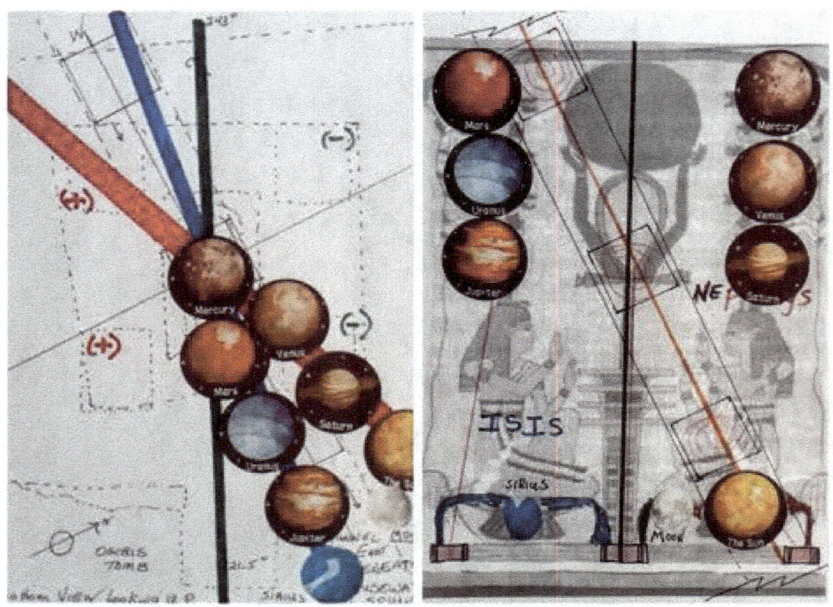

Pyramid systems are **impeccable in both design and function**, unlike anything else on the planet. It is time to **awaken to their potential** and replicate them as best we can—for the **benefit of the Earth and humanity**. However, I believe that certain agendas exist today to discourage this mindset.

The following image is the **second of four Papyrus hieroglyphs** that I have translated differently. These hieroglyphs bear a striking resemblance to the **Papyrus of Ani**.

Below is an image of **Papyrus of Hunefer, Plate #1, dating back to 1275 BC**—created by different artists in different eras, yet depicting the same enduring theme.

THE GIZA PARK PROTOCOL HOW TO START THE SUN

It is easy to see the resemblance to the **Papyrus of Ani**, dated **1250 BC**.

However, two plates from the **Papyrus of Quenna—Plates #1 and #3**—are the earliest, dating back to **1550 BC**, and they also align with my theories.

Here is one of them:

THE GIZA PARK PROTOCOL HOW TO START THE SUN

As you can see, the **Quenna hieroglyph** is much more **primitive and crude**, yet it still conveys the same message.

By reinterpreting the meaning of **four major hieroglyphs**, we must ask: **How many other translations have been influenced by these earlier interpretations?** Could it be **hundreds or even thousands**?

I hope my insights into the **design and function of Pyramid Systems** have offered you a fresh perspective.

At best, this volume serves as an **overview of highly complex systems**—mysteries that have perplexed humankind for centuries.

THE GIZA PARK PROTOCOL HOW TO START THE SUN

The overwhelming **physical evidence** suggests that a **Level One Civilization** was responsible for these **advanced Pyramid Systems**.

This text is only the **first volume** of what could become an **extensive body of research**. As I begin **Volume Two**, I look forward to **collaborations, contributions, and critical insights** from new researchers.

Chapter Fifteen:
Summation overall

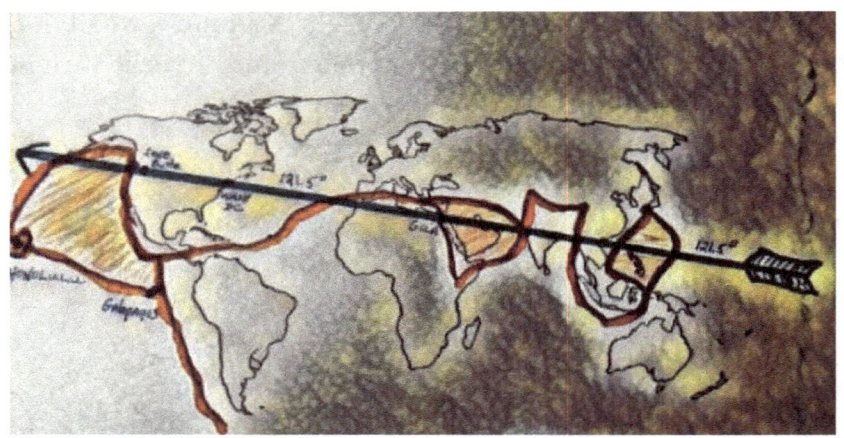

The well-engineered and precisely designed Weather and Climate Modification Systems operated in **two distinct phases**. Both were **perfectly timed** to end the Ice Age without flooding entire continents or destroying existing habitats.

This system harnessed the **natural energies of the Earth and the universe**, entangling them through an **alchemy perfectly suited for human habitation at a quantum level**.

By synchronizing with **natural cyclic events**, this method **capitalized on existing strengths** to generate a **rapid, controlled abundance for humankind**, enabling greater **freedom of movement and exploration**.

Keep in mind that the **Ice Age lasted between 2.4 and 3 million years**, making the Earth a **harsh and inhospitable** place for human life.

What follows is the result of my research—I am offering an explanation of **how it all occurred**.

THE GIZA PARK PROTOCOL HOW TO START THE SUN

The collage image below illustrates the **location of the 2-mile-thick ice sheet** that once covered the **northern regions of the North American continent**.

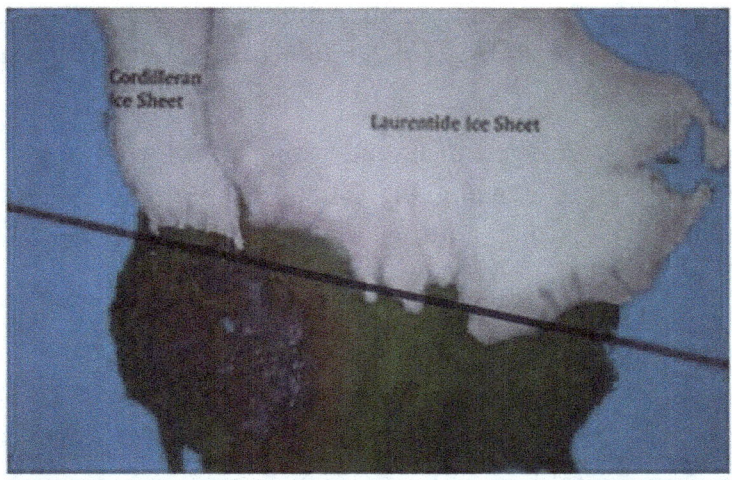

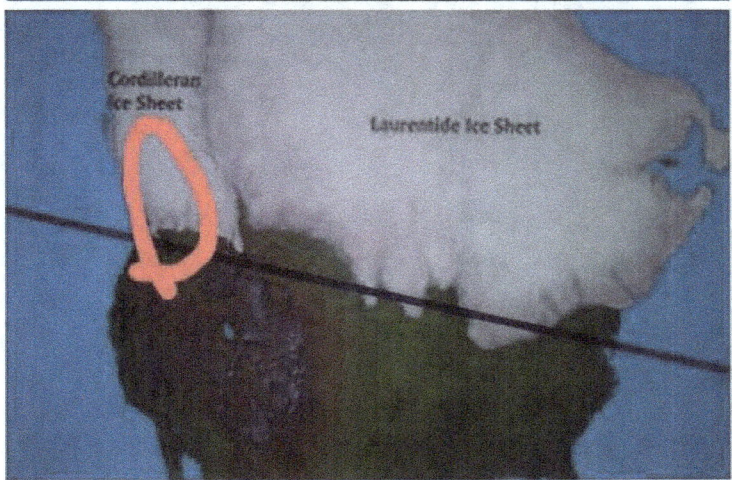

The **black line** marks the location of the **Osiris Line**, which is particularly intriguing because the **thick ice sheet** never extended beyond this boundary.

THE GIZA PARK PROTOCOL HOW TO START THE SUN

In the **bottom image**, the **red circle** highlights a region containing **44 Tuya subglacial volcanoes**, all of which **formed beneath the immense weight of the ice sheet**.

Once again, the **two-phase operation** was **brilliantly designed and executed** to bring the **Ice Age to a controlled and precisely timed conclusion**.

Below is the **image I provided** to illustrate the **cyclic patterns** involved. I am referencing the **24,000-year Precession of the Equinoxes cycle**, which I introduced earlier.

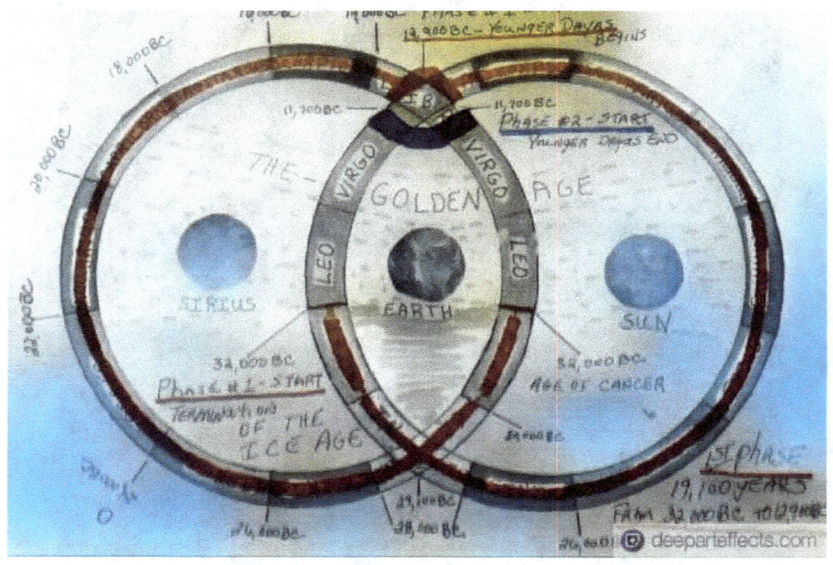

Phase #1

The first phase began in **32,000 BC**, during the **Age of Cancer**, a **water sign**. This initial cycle lasted **19,100 years**, concluding in **12,900 BC**, during the **Age of Libra**, the sign of **balance and justice**—marking the onset of the **Younger Dryas period**.

This **well-planned event** was intentionally designed to **control surface glacial melting** and **prevent extreme flooding** across all continents, particularly **North America**.

THE GIZA PARK PROTOCOL HOW TO START THE SUN

Modern scientists estimate that the **end of the Ice Age** began around **14,500 BC**, near the end of the **Age of Scorpio**, another **water sign**. Their conclusions are based on **a single catastrophic event** of immense proportions:

The **Bonneville Flood**—the **largest known catastrophe on the North American continent**.

I firmly believe that the **Bonneville Flood** provides **overwhelming physical support for my theories**.

First, I am convinced that the majority of the water responsible for this **great flood** was **subglacial meltwater** released from beneath the **two-mile-thick ice sheet**.

This also perfectly explains the existence of the **Tuya subglacial volcanoes** I referenced earlier.

The following image highlights the **geography of the Pacific Northwest Bonneville floodplain**, which is strikingly close to the **44 Tuya subglacial volcanoes**.

THE GIZA PARK PROTOCOL HOW TO START THE SUN

During this **first phase** of the **Weather and Climate Modification Systems (WCMS)** operation, the **entire Pacific Northwest**—along with much of the globe—was buried beneath a **two-mile-thick ice cap** that encircled the planet, making human survival nearly impossible.

As we know, **volcanoes cannot function** when trapped within such a **massive frozen barrier**. **Magma**, the core ingredient of lava, requires **at least 70% water** to flow and move. This meant that **hundreds, if not thousands, of volcanoes** across the world were **completely inactive**.

So, what was needed to **end the Ice Age**? **Heat.** And with volcanoes—the Earth's **primary surface heat source**—essentially **"broken,"** an alternative method was required.

The Giza Connection: A Lost Technology

My theories present a compelling opportunity to **rethink how we view our planet**, and one of the first key insights lies in **understanding volcanoes differently**.

The **WCMS start-up sequence** began with **Egypt's most advanced and diverse system**—the **Giza Industrial Park Systems**. At its core was the **Plasma EMF Scaler Cyclotron**, known today as **Osiris Tomb**.

This **powerful energy device** lies **108 feet beneath the desert sands**, positioned between the **Great Pyramid** and the **Second Pyramid**—which I refer to as the **Ionic Plasma Reflux Pyramid**.

The Osiris Tomb functions as an **ancient scaler energy system**, comparable to today's **CERN**, but vastly more advanced in purpose. It generates **scalar energy faster than the speed of light**, transmitting it deep into the Earth's surface.

After extensive research, I traced the **Osiris energy path** as it traveled from Egypt, across Africa, and over the Atlantic Ocean. My first key discovery brought me to **Ocean City, Maryland**, where I made landfall after departing from the **Atlantic shores of Casablanca**.

From this point, I traveled **exactly 111 miles inland**, arriving at a significant landmark—the **Washington Monument**. Its rich **symbolism** and direct connections to **Egypt, Osiris Tomb, and Israel** made it the first major **energy-linked location** I recognized.

Tracing The Energy to The West Coast

Continuing my journey across America, I made a **groundbreaking discovery** in **Washington State**, just south of **Mount St. Helens**.

I realized that the **primary function of Egypt's pyramid systems** was to **activate subglacial volcanoes beneath the ice**—a brilliant strategy to **melt ice from below** and **generate the water needed for magma formation**.

After following this **energy anomaly** from **108 feet below the ground in Egypt**, across the **African continent**, the **Atlantic Ocean**, and **North America**, I arrived at a truly **rare geological site**:

Lone Butte—A Hidden Key to the Ice Age Melt

Lone Butte, a unique **Tuya volcano**, became my next major discovery. Affectionately known as **"Flat Top" or "Tabletop" volcano**, its distinct shape tells a fascinating story.

Lone Butte **erupted beneath a two-mile-thick ice sheet**, and the **immense weight of the ice** acted as a **pressure barrier**. The **plasma-scaler energy** from the **Osiris Tomb** in Egypt **superheated magma beneath the glacier**, forcing **high-pressure molten rock** upward, melting rivers of ice from below.

But Lone Butte is just the beginning.

Two additional **Tuya volcanoes**, similar in nature, are located just **south of Mount Hood**. These **three rare Tuya volcanoes** were, in my theory, **directly activated by the scalar energies from Osiris Tomb**, initiating a **controlled melting of the Ice Age glaciers from below**.

This **genius-level engineering** remains one of the most **brilliantly designed climate control systems** ever conceived.

THE GIZA PARK PROTOCOL HOW TO START THE SUN

The Osiris Line and the Ring of Fire

The **Osiris energy pathway** continues, ultimately connecting with **Earth's most powerful volcanic region**—the **Ring of Fire.**

Stretching an astonishing **40,000 km in length and 500 km in width**, the **Ring of Fire** contains **between 750 and 915 active and dormant volcanoes**—accounting for nearly **two-thirds of all volcanoes on Earth.**

This extraordinary alignment suggests that the **ancient energy grid** designed by past civilizations was far more **advanced and intentional** than we have ever imagined.

If we turn our gaze from Lone Butte to the northeast, crossing the Canadian border into British Columbia, we discover a cluster of 41 Tuya volcanoes at the heart of the "Ring of Fire." What a remarkable discovery! Now we understand where the water originated that caused the catastrophic Bonneville Flood.

Using this data, I concluded that it took 17,500 years of diurnal cyclic operations by at least three Pyramidal Systems to perfect and complete the project as designed.

The recorded date of the Bonneville Flood, 14,500 BC, was the result of direct activity from 44 Tuya volcanoes clustered in the higher elevations of British Columbia, Canada, and the Pacific Northwest region of the United States.

In 14,500 BC, following the Bonneville Flood, the necessary process of thawing ice into liquid water was required to awaken dormant and frozen magma fields. Since the controlled melting of ice had been commissioned, these newly active surface volcanoes operated in conjunction with the Weather and Climate Modification Systems (WCMS) until Phase 1 ended in 12,900 BC.

Thus, the Younger Dryas period began, slowing the rapid melting of ice and preventing a flood event that could have overwhelmed continents and disrupted ocean currents. Scientists estimate that global water levels rose by approximately 300 feet during this time.

The Younger Dryas lasted 1,200 years, beginning with the system shutdown in 12,900 BC.

Phase 2 began in 11,700 BC, during the Age of Libra. It is important to note that the diurnal cyclic WCMS was halted in Libra, and it was also in Libra that the Weather and Climate Modification Systems were reactivated to restore balance.

With this, the Younger Dryas period ended, and the systems remained operational from 11,700 BC until 3,000 BC—a span of 8,700 years. In 3,000 BC, the rains ceased, ushering in a dry period in Egypt, which ultimately contributed to the decline of the Egyptian dynasties.

By 3,000 BC, the Age of Taurus—symbolising Earth—had begun. At this point, the Weather and Climate Modification Systems were shut down, and further operations were unlikely to continue as technological advancements were expected for Earth's future inhabitants.

Weather and Climate Modification Systems (WCMS)

- **32,000 BC (Age of Cancer):** Initial activation of all systems, including the Egyptian Pyramidal Systems, the Bosnian Pyramid Systems, and the Teotihuacán Pyramid Systems in Mexico. Additionally, Angkor Wat in Cambodia and the Chinese pyramids are suspected of involvement.

- These systems operated for **19,100 years**, generating sufficient heat energy to encourage the melting of ice from beneath the surface, contributing to the end of the Ice Age.

- **11,700 BC (Age of Libra):** Phase 2 began, marking the end of the planned cooling period and the immediate reactivation of the WCMS.

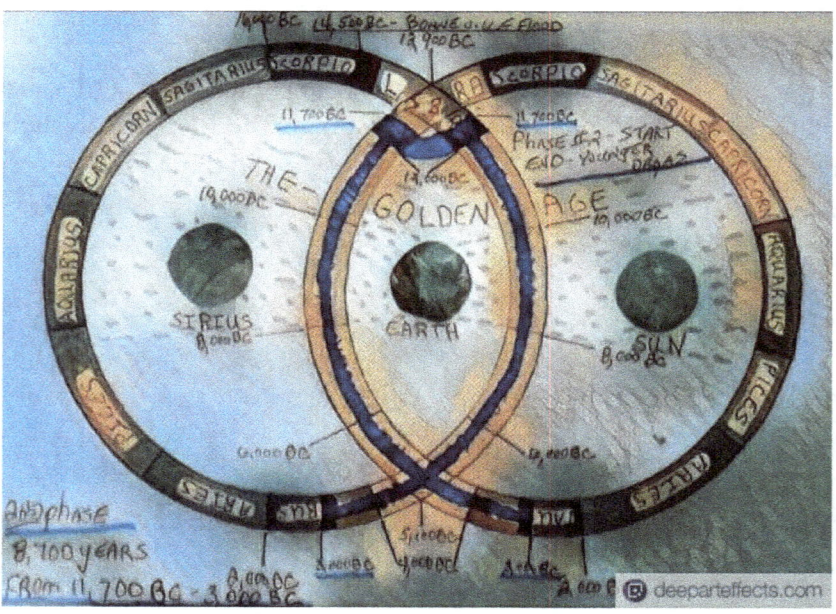

The **Golden Age** began just over 2,000 years later, in the **Age of Leo**, which commenced in **10,000 BC**. Whatever systems orchestrated this period of abundance on Earth were impeccably timed, precisely manipulated, and flawlessly executed.

If an earthling had designed this magnificent work of art, they would have been awarded the **Nobel Prize in Science**.

As we can see, the **second phase** of the **Weather and Climate Modification Systems (WCMS)** concluded in the **Age of Taurus**, around **3,000 BC**, when the rainy season began to diminish. This transition ultimately led to the onset of the **Dry Period**, which emerged **500 years later, in 2,500 BC**.

The **second phase** of the WCMS operation lasted **8,700 years**.

With precise execution and perfect timing, the total duration of both operational phases amounted to **27,800 years**—successfully bringing an end to Earth's **2.4-million-year Ice Prison** and allowing the human species to thrive. The termination of the Ice Age stands as a monumental achievement.

THE GIZA PARK PROTOCOL HOW TO START THE SUN

A Personal Reflection

As I continue writing and preparing the next volume—where I will explain much more—I extend my deepest gratitude to every reader who has made it to this page. Thank you for listening. Hopefully, pyramid research will evolve to a **higher, more intelligent dimension**.

To conclude this **first volume**, I wish to reference **Hermes Trismegistus**, quoting his best-translated insights from the **Emerald Tablets**:

"If we can keep an area of the Earth protected from the burning rays of the Sun by constructing a structure capable of harnessing all subtle and low-frequency energies..."

"By building upon natural strengths in an increasing and positive direction, strength adds to strength... This combination of forces can halt all interference or be manipulated by any other energy fields."

"The power gained is of such magnitude that it can penetrate and conquer all things—whether through the subtlest forms of air and water or the densest, heaviest mass."

Hermes further declared:

"This method of alchemy is how the world was made—through the combination of matter and all known energies within this marvellous system. Remember, these benefits are assured every time you follow this design."

His **closing statement** is particularly revealing—he speaks of the **Pyramid Systems**:

"My name is Hermes Trismegistus, and this written communication is proof that what I say is true... My knowledge and

THE GIZA PARK PROTOCOL HOW TO START THE SUN

understanding of the philosophy of the Principles of the Universe..."

"I declare this Solar work project complete, tested, and ready for operation."

A Final Thought

Be unconquerable. Seek truth and science with integrity and reason.

www.ingramcontent.com/pod-product-compliance
Lightning Source LLC
Chambersburg PA
CBHW052007070526
44584CB00016B/1647